Springer Aerospace Technology

Series Editors

Sergio De Rosa, DII, University of Naples Federico II, NAPOLI, Italy

Yao Zheng, School of Aeronautics and Astronautics, Zhejiang University, Hangzhou, Zhejiang, China

The series explores the technology and the science related to the aircraft and spacecraft including concept, design, assembly, control and maintenance. The topics cover aircraft, missiles, space vehicles, aircraft engines and propulsion units. The volumes of the series present the fundamentals, the applications and the advances in all the fields related to aerospace engineering, including:

- structural analysis,
- aerodynamics,
- aeroelasticity,
- aeroacoustics,
- flight mechanics and dynamics,
- orbital maneuvers,
- avionics,
- systems design,
- materials technology,
- launch technology,
- payload and satellite technology,
- space industry, medicine and biology.

The series' scope includes monographs, professional books, advanced textbooks, as well as selected contributions from specialized conferences and workshops.

The volumes of the series are single-blind peer-reviewed.

To submit a proposal or request further information, please contact:

Mr. Pierpaolo Riva at pierpaolo.riva@springer.com (Europe and Americas) Mr. Mengchu Huang at mengchu.huang@springer.com (China)

The series is indexed in Scopus and Compendex

More information about this series at http://www.springer.com/series/8613

Vadim Serafimovich Kalashnikov ·
Maxim Yurievich Ponomarev ·
Oleg Yurievich Platonov ·
Victor Vasilievich Shubnikov · Mark Ilyich Rivkin ·
Artem Yurievich Shatrakov ·
Yury Grigorievich Shatrakov ·
Oleg Ivanovich Zavalishin

Near-Field Antenna Measurements

Calculations and Facility Design

Springer

Vadim Serafimovich Kalashnikov
Saint Petersburg, Russia

Maxim Yurievich Ponomarev
Saint Petersburg, Russia

Oleg Yurievich Platonov
Saint Petersburg, Russia

Victor Vasilievich Shubnikov
Saint Petersburg, Russia

Mark Ilyich Rivkin
Saint Petersburg, Russia

Artem Yurievich Shatrakov
Moscow, Russia

Yury Grigorievich Shatrakov
Saint Petersburg, Russia

Oleg Ivanovich Zavalishin
Moscow, Russia

ISSN 1869-1730 ISSN 1869-1749 (electronic)
Springer Aerospace Technology
ISBN 978-981-33-6435-6 ISBN 978-981-33-6436-3 (eBook)
https://doi.org/10.1007/978-981-33-6436-3

This Springer imprint is published by the registered company Springer Nature Singapore Pte Ltd.
The registered company address is: 152 Beach Road, #21-01/04 Gateway East, Singapore 189721, Singapore

Preface

Questions of determination of antenna parameters on the basis of measurement results of field amplitude–phase distribution on scanning surface in the near field of these antennas are reviewed. The study contains electrodynamic basis of near-field measurements and ready to use formulas to determine antenna parameters, based on measurement of amplitude–phase field distribution on flat and spherical scanning surfaces. Recommendations on how to write fast and efficient MATLAB (Octave) code, enabling to implement these formulas, are given. Measurement error estimation is carried out. The description of purpose designed facilities for measurement of amplitude–phase distribution of the field on flat scanning surfaces in the near-field zone of the studied antennas is given. The questions related to laboratory and flight tests of antennas are considered.

The book is intended for engineers and highly qualified specialists, whose activities are related to experimental testing of radio characteristics of complex antenna systems, and can also be used as a textbook for senior students in the field of "radioelectronics" and "radiophysics".

Saint Petersburg, Russia	Vadim Serafimovich Kalashnikov
Saint Petersburg, Russia	Maxim Yurievich Ponomarev
Saint Petersburg, Russia	Oleg Yurievich Platonov
Saint Petersburg, Russia	Victor Vasilievich Shubnikov
Saint Petersburg, Russia	Mark Ilyich Rivkin
Moscow, Russia	Artem Yurievich Shatrakov
Saint Petersburg, Russia	Yury Grigorievich Shatrakov
Moscow, Russia	Oleg Ivanovich Zavalishin
2019	

Introduction

Development of information technologies and mathematical modeling methods enabled scientists and specialists to use the obtained results to expand the discipline related to evaluation of antenna characteristics, based on measurements in the near field. Development of antennas for ground-based, ship-based, space-based, and airborne radio-electronic systems is a complex and expensive process. Testing of antenna parameters in the course of their development takes a long time. Developers have to constantly evaluate these parameters, when changing the antenna design due to the need to take into account the impact of carrier object elements and external destabilizing effects on these parameters. Implementation of the technologies related to antenna parameter evaluation on the basis of characteristics, measured in the near field, in practical activities of R&D centers and industrial enterprises not only increases the competitiveness of developed antenna products, but also allows to win new international markets. Using the near-field facility, professionals are given the opportunity to practically assess the impact of carrier object elements on alteration of antenna characteristics, thus allowing not only to consider the diffraction problems, but also to establish on-board antenna parameters, taking into account provision of stable radio communication link.

The monograph is based on the results of the authors' research activities in the course of development of antennas and antenna systems for a wide range of carrier objects (aircraft, mobile ground systems, etc.). One of the essential features of this monograph is that the proposed presentation of the issues is focused on both scientists and designers of radio engineering systems. The monograph includes an introduction, list of symbols and abbreviations, and six chapters. Reference literature is given at the end of each chapter.

Chapter 1 is dedicated to the definition and classification of main radio characteristics of antennas.

Chapter 2 provides a brief overview of methods of radio characteristic measurement of antennas.

Chapter 3 is dedicated to antenna characteristic calculation on the basis of measurement results of field amplitude–phase distribution on plane scanning surface.

Chapter 4 is dedicated to the study of characteristics of the planar near-field facility.

Chapter 5 is dedicated to the calculation of antenna characteristics on the basis of measurement of amplitude–phase field distribution on spherical scanning surface.

Chapter 6 is dedicated to mechanical, climatic, and flight antenna testing.

The monograph was developed by the team of contributors, consisting of:

Vadim Serafimovich Kalashnikov
Oleg Yurievich Platonov
Maxim Yurievich Ponomarev
Mark Ilyich Rivkin
Artem Yurievich Shatrakov
Yury Grigorievich Shatrakov
Victor Vasilievich Shubnikov
Oleg Ivanovich Zavalishin

This monograph will be useful for radio specialists and for radiophysicists, whose activities are associated with antenna development.

Contents

Symbols and Abbreviations

AUT	Antenna under test
D	Directivity
DFT	Discrete Fourier transform
EM	Electromagnetic
FFT	Fast Fourier transform
G	Gain
G_R	Gain realized
Γ	Complex reflection coefficient
LAN	Local area network
L_{RET}	Return loss
PWS	Planar wave spectrum
RF	Radio frequency
RP	Radiation pattern
3D RP	Three-dimensional radiation pattern
UHF	Ultrahigh frequency
VSWR	Voltage standing wave ratio
η	Efficiency
Z_A	Complex antenna impedance

Chapter 1
Radio Characteristics of Antennas

1.1 The Role of Antennas in Radio Systems

Any radio line designed to transmit information or energy through free space contains transmitting and receiving antennas.

A simplified block diagram of such a radio line is shown in Fig. 1.1. It consists of a transmitter (1), a transmitting antenna (3), a transmission line segment connecting the output of the transmitter to the input of the transmitting antenna (2), a free space in which electromagnetic waves propagate (4), a receiving antenna (5), a receiver (7), and a transmission line segment connecting the output of the receiving antenna to the input of the receiver (6).

The main tasks solved by transmitting antennas are the following:

(1) conversion of the energy of high-frequency currents excited by the transmitter at the antenna input into the energy of electromagnetic waves emitted by the antenna into free space;
(2) energy beaming by excited electromagnetic waves, i.e., ensuring the required dependence of the power flux density of these waves on the direction in which they leave the antenna.

The main tasks solved by receiving antennas are the following:

(1) conversion of the energy of electromagnetic waves incident on the antenna into the energy of high-frequency currents excited by these waves at the antenna output;
(2) directed reception of energy of electromagnetic waves incident on the antenna, i.e., ensuring the required dependence of the electromotive force induced at the output terminals of the antenna on the direction from which these waves incident on it.

Summarizing the above, one can say that the antennas match the transmitter and receiver with the free space and perform directed transmission and directed reception of energy carried by electromagnetic waves.

V. S. Kalashnikov et al., *Near-Field Antenna Measurements*, Springer Aerospace Technology, https://doi.org/10.1007/978-981-33-6436-3_1

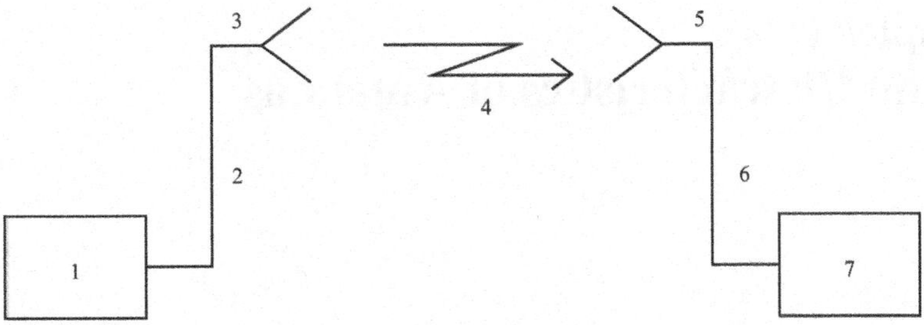

Fig. 1.1 Block diagram of the radio line

Transmitting and receiving antennas increase the efficiency of the radio line, allowing for a fixed transmitter power and a fixed receiver sensitivity to adjust the maximum range of its stable operation.

The quality of the transmitting and receiving antennas is assessed by a number of quantitative characteristics that can be calculated and measured. These characteristics are called "radio antenna characteristics," or simply "antenna parameters."

The introduction of these parameters, for which the calculation formulas and measurement methods are defined, provides a compact form for the presentation of a technical task for the development of antennas and a justified selection of antenna types intended to solve specific technical problems.

Antenna parameters associated with the energy of high-frequency currents excited in the antenna by the transmitter (when operating in the transmitting mode) or incident electromagnetic waves (when operating in the receiving mode) are called internal antenna parameters.

The antenna parameters associated with the energy of emitted (when operating in the transmitting mode) or received (when operating in the receiving mode) electromagnetic waves are called external antenna parameters.

1.2 The Structure of Electromagnetic Fields Excited by the Antenna

1.2.1 Types of Variable Electromagnetic Fields Excited by the Antenna in the Surrounding Space

The antenna excites two types of variable electromagnetic fields in the surrounding space, propagating and non-propagating ones.

Propagating variable electromagnetic fields take energy from the antenna and irrevocably carry it into the surrounding space. The period average value of the power flux density carried by these fields is always greater than zero. These fields are called radiation fields or electromagnetic waves.

Non-propagating alternating electric and magnetic fields are concentrated near the antenna and only exchange energy with each other and with the antenna. The average value of the power flux density carried by these fields over a period is always zero. These fields are called reactive fields or induction fields.

The main parameters of an electromagnetic (EM) field are strength, intensity, power, phase, polarization, and frequency. Let us recall these parameters [1–3].

EM field strength is a value that characterizes the amplitude of the vector of the electric or magnetic component of the EM field at a certain point in space.

EM field intensity is the square of the EM field strength.

EM field power is the amount of energy of the EM field that passed through some surface per unit of time.

The power flux density of the EM field is the power of the EM field that passed through a unit area in the direction perpendicular to this area.

The phase of an EM field is the angle of the phase shift between the oscillations of the EM field at a given point and at a conventional point of origin at the same time.

Wavefront (phase front) is the surface of equal phases of the electromagnetic field.

The polarization of the EM field is a characteristic of the EM field that determines the behavior of the electric field intensity vector in the transverse plane (perpendicular to the direction of propagation of the EM field) for a period of time equal to the period of oscillations.

The frequency of EM field is the frequency of oscillations of its electric or magnetic component.

1.2.2 Near-Field and Far-Field Regions of the Antenna

Depending on the structure of the electromagnetic field excited by the antenna, the space in which it exists can be divided into several regions (zones). Let us examine the location of these zones by the example of an aperture antenna focused at infinity.

Adjacent to the antenna surface is the near-field region of the antenna, which is divided into two sub-regions: the reactive near-field region and the radiative near-field region.

The area directly adjacent to the antenna surface is the reactive near-field region, in which the strength of these fields significantly exceeds the strength of the radiation fields. However, due to their physical nature, their intensity decreases inversely proportional to the second or even third degree of distance from the antenna surface (depending on the type of antenna). At the same time, the intensity of the radiation fields in this region is almost independent of the distance to the antenna surface. The conventional outer boundary of the zone of reactive fields, after which they can practically be neglected in comparison with radiation fields, is at a distance λ/π from the antenna surface.

The first part of the radiative near-field region, the aperture zone, begins from the reactive near-field region and extends to a distance of approximately $0.62 \cdot D \cdot (D/\lambda)^{0.5}$

(where D is the maximum linear size of the antenna and λ is its operating wavelength). In this area, the reactive fields are negligible, the front of the propagating electromagnetic wave is quasi-plane (almost non-divergent projecting ray), the field strength is almost independent of the distance to the antenna, and the distribution of the field amplitudes replicates the form of the distribution of amplitudes in the antenna aperture.

The second part of the radiative near-field region, the Fresnel region, is the area of space with conventional boundaries from $0.62 \cdot D(D/\lambda)^{0.5}$ to $2D^2/\lambda$. In the Fresnel region, the relative angular distribution of the field amplitude (radiation pattern) starts to form, but the shape of this distribution depends on the distance to the antenna. There are two reasons for this: first, the phase relationships between the fields from different antenna elements change with distance, and second, the ratio of field amplitudes from different antenna elements also changes with distance. As the observation point moves away from the antenna (along the radius), the amplitude of the field first oscillates and then decays steadily. In the infinite distance limit, this attenuation is inversely proportional to the first power of the distance. In addition, as the observation point moves further away, the relative phase and amplitude relationships between the fields from the individual antenna elements asymptotically approach fixed values, and the angular distribution of the field no longer depends on the distance. Although this position, strictly speaking, is achieved only when the observation point is moved to infinity, with certain assumptions, it is possible to find the outer boundary of the Fresnel zone. For most aperture antennas, this boundary is determined by the distance $2D^2/\lambda$.

The far-field region of the antenna extends from a distance of $2D^2/\lambda$ to infinity. In the far zone, the ratio of amplitudes and phases of the field from different parts of the antenna does not depend on the distance to the antenna and is determined only by the angular position of the observation point. When the observation point moves around a directional antenna, a stable interference pattern is observed, which is an alternation of maxima and minima of the field strength. This pattern is an amplitude radiation pattern (RP) of the antenna (see clause 1.4). The section of the RP between adjacent minima is called a lobe of the RP. Usually, a distinction is made between the main and the side lobes of the RP, and they are characterized by their maximum field value (absolute or relative) and angular width at a given level (usually at levels -3, -10 dB, or at the level of the minima that define the lobe; ideally, this level is $-\infty$ dB). The angular distribution of the field in the far region (the shape of the radiation pattern) does not depend on the distance to the antenna, the amplitude of the field descending proportionally to the first degree of the distance from the antenna, and the vectors $\boldsymbol{E}$ and $\boldsymbol{H}$ being mutually perpendicular and located in a transverse plane with respect to the direction of propagation. The phase front of the electromagnetic wave in the far region is spherical (within small space angles it is flat).

The internal parameters of the transmitting antennas are related to the reactive and radiated electromagnetic fields excited by the antenna.

The external parameters of the transmitting antennas are related only to the radiated electromagnetic fields (electromagnetic waves) excited by the antenna.

1.3 Internal Parameters of Transmitting Antennas

The internal parameters of the transmitting antenna characterize it as an energy consumer and assess the degree of matching of the antenna input with the transmission line connecting it to the transmitter (the antenna input is understood as the terminals to which the transmission line between the transmitter and the antenna is connected). The internal parameters of the transmitting antenna include the complex input impedance, the efficiency (η), the reflection coefficient (Γ), the standing wave ratio (VSWR), and return loss (L_{RET}).

1.3.1 The Balance of Active and Reactive Power in the Antenna

The active power coming from the transmitter to the antenna is partially reflected from its input and partially passes into the antenna. The active power that passes is partially spent on active losses in the antenna, and the remaining part is transferred to the electromagnetic waves excited by the antenna and radiated into the surrounding space.

Let us introduce the following notation:

P_{TRM} is the power from the transmitter arrived to the antenna input.
P_{REF} is the power reflected from antenna input.
P_A is the power accepted by the antenna.
P_L is the power of active losses caused by the flow of the conduction and displacement currents excited by the electromagnetic field of the antenna in the metal and dielectric elements of the antenna structure.
P_R is the power transferred by the antenna to the electromagnetic waves excited by it (radiated power).

A graphical representation of the active power balance at the input of the transmitting antenna is shown in Fig. 1.2.

The power of P_{TRM} is divided into P_A and P_{REF} ($P_{TRM} = P_A + P_{REF}$).

The power of P_A is divided into P_L and P_R ($P_A = P_L + P_R$), hence:

$$P_{TRM} = P_L + P_{REF} + P_R \tag{1.1}$$

The presence or absence of reactive power at the antenna input is related to the balance of energy stored by reactive electric and magnetic fields excited in the near zone of the antenna. Over time, these fields exchange energy, completing a full cycle of exchange in a period of time equal to the period of excited oscillations. If the maximum amount of energy that electric and magnetic reactive fields can store is not equal to each other, the source (transmitter) is also involved in the exchange of

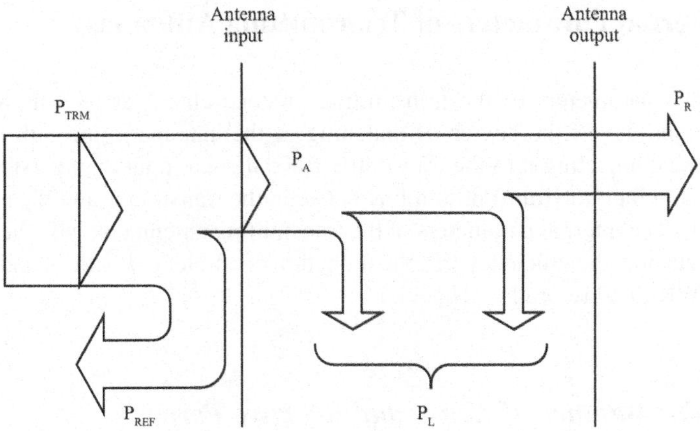

Fig. 1.2 Active power balance in the transmitting antenna

energy, and reactive power appears at the antenna input. If the energies stored are equal, the reactive power at antenna input is equal to zero.

Let us introduce the following notation:

P^{E} is the maximum power that can be stored by the electric induction field in the near zone of the antenna.
P^{M} is the maximum power that can be stored by the magnetic induction field in the near zone of the antenna.
P^{REACT} is the reactive power at the antenna input.

The relationship between these powers is determined by the following equation:

$$P^{\mathrm{REACT}} = \left|P^{\mathrm{E}} - P^{\mathrm{M}}\right| \tag{1.2}$$

If antenna is designed in such a way that P^{E} P^{M}, then P^{REACT} 0.
If antenna is designed in such a way that $P^{\mathrm{E}} = P^{\mathrm{M}}$, then $P^{\mathrm{REACT}} = 0$.

1.3.2 Input Impedance of the Antenna

The equivalent circuit of the transmitting antenna as the energy consumer of the transmitter is presented in Fig. 1.3. The antenna is a complex load (Z_{A}) having active and reactive components, which is connected to the transmitter by a lossless long line having a purely active wave impedance Z_{o}.

$$Z_{\mathrm{A}} = R_{\mathrm{A}} + jX_{\mathrm{A}} \tag{1.3}$$

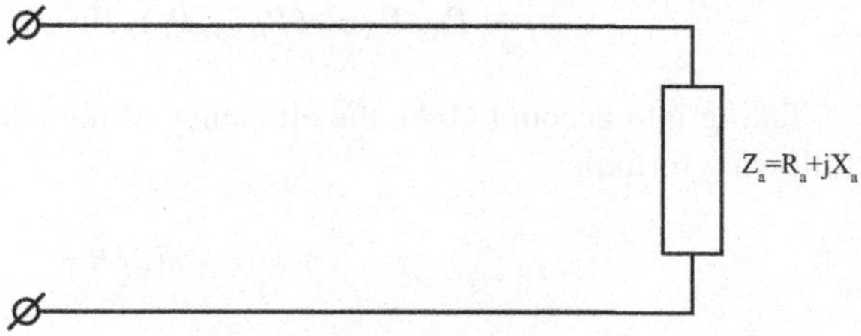

Fig. 1.3 Equivalent circuit of a transmitting antenna

Z_A resistance is called the input impedance of the antenna, and its components are the active (R_A) and reactive (X_A) parts of the input impedance of the antenna.

The physical meaning of the R_A and X_A parameters can be understood by analyzing the balance of active and reactive power in the antenna. If one connect the powers of P_A, P_R and P_L with the current at the antenna input (I_{IN}), then their characteristics can be described formally with active resistance R_A, R_R and R_L, where R_A is the active component of input impedance of the antenna, R_R is the radiation resistance of the antenna, and R_L is the resistance of active losses in the antenna:

$$P_A = 0.5 \cdot I_{IN}^2 \cdot R_A;\ P_R = 0.5 \cdot I_{IN}^2 \cdot R_R;\ P_L = 0.5 \cdot I_{IN}^2 \cdot R_L; \qquad (1.4)$$

It is obvious that the resistances R_A, R_R and R_L are connected by the following relation:

$$R_A = R_R + R_L; \qquad (1.5)$$

The formulas for calculating the R_A, R_R, and R_L are derived for the study of specific types of antennas.

To connect the P^{REACT} power with the current at the input of the antenna, it is possible to formally introduce reactive resistance X_A, which is a reactive component of input resistance of the antenna:

$$P^{REACT} = 0.5 \cdot I_{IN}^2 \cdot |X_A| \qquad (1.6)$$

Thus, the reactive component of the antenna input impedance X_A is a measure of the reactive power at the antenna input.

The formulas for calculating the X_A parameter are derived for the study of specific types of antennas.

1.3.3 *The Efficiency of the Antenna*

Antenna efficiency (η) describes the relation between P_A, P_R, and P_L powers (see Sect. 1.3.1).

$$\eta = P_R/P_A = (P_A - P_L)/P_A = 1 - P_L/P_A \quad (1.7)$$

Taking into account (1.4), the efficiency of an antenna can be calculated by the following formula:

$$\eta = 1 - R_L/R_A \quad (1.8)$$

It should be emphasized that the efficiency of the antenna does not depend on the degree of matching of its input (the power reflected from the antenna input (P_{REF}) is not taken into account).

1.3.4 The Parameters that Characterize the Match of the Antenna Input

The degree of matching of the antenna input with the output of the transmission line connecting it to the source (transmitter) is accessed with the use of one of the following parameters [2, 3]:

1. the modulus of the complex voltage reflection coefficient at the input of the antenna ($|\Gamma|$);
2. voltage standing wave ratio (VSWR) at the antenna input;
3. Return loss at the antenna input (L_{RET} [dB]).

The parameters $|\Gamma|$, VSWR and L_{RET} can be expressed in terms of the active power reflected from the antenna input (P_{REF}) and the active power coming to the antenna input from the transmitter (P_{TRM}):

$$|\Gamma| = \sqrt{P_{REF}}/\sqrt{P_{TRM}}; \quad (1.9)$$

$$\text{VSWR} = (\sqrt{P_{TRM}} + \sqrt{P_{REF}})/(\sqrt{P_{TRM}} - \sqrt{P_{REF}}); \quad (1.10)$$

$$L_{RET}[\text{dB}] = 10\lg(P_{REF}/P_{TRM}) \quad (1.11)$$

It is obvious that the parameters $|\Gamma|$, VSWR, and L_{RET} have simple mathematical relations to each other.

$$|\Gamma| = (\text{VSWR} - 1)/(\text{VSWR} + 1); \quad (1.12)$$

$$\text{VSWR} = (1 + |\Gamma|)/(1 - |\Gamma|); \quad (1.13)$$

$$L_{RET}[\text{dB}] = 10\lg|\Gamma|^2 = 20\lg|\Gamma|. \quad (1.14)$$

In terms of the active power balance at the antenna input, its full matching takes place when the following conditions are met:

$$P_{\mathrm{REF}} = 0,\ P_{\mathrm{A}} = P_{\mathrm{TRM}} \tag{1.15}$$

In this case,

$$|\Gamma| = 0,\ \mathrm{VSWR} = 1,\ L_{\mathrm{RET}} = -\infty. \tag{1.16}$$

Let us find out at what values of the active and reactive components of the input impedance of the antenna its full matching occurs.

For this purpose, it is necessary to use the formula for the calculation of the complex reflection coefficient of a lossless long line with wave resistance Z_o loaded on the complex resistance Z_A.

$$\Gamma = (Z_{\mathrm{A}} - Z_0)/(Z_{\mathrm{A}} + Z_0) = ((R_{\mathrm{A}} - Z_0) + jX_{\mathrm{A}})/((R_{\mathrm{A}} + Z_0) + jX_{\mathrm{A}}) \tag{1.17}$$

hence

$$|\Gamma|^2 = ((R_{\mathrm{A}} - Z_0)^2 + X_{\mathrm{A}}^2)/((R_{\mathrm{A}} + Z_0)^2 + X_{\mathrm{A}}^2) \tag{1.18}$$

When the antenna is fully matched, $|\Gamma|$ must be zero. From Eq. (1.17), it follows that $|\Gamma| = 0$ if the following conditions are met:

$$R_{\mathrm{A}} = Z_0,\ X_{\mathrm{A}} = 0. \tag{1.19}$$

Thus, to achieve full matching of the antenna, it is necessary that the active component of its input resistance is equal to the wave resistance of the transmission line through which the antenna receives power from the transmitter, and the reactive component of the input resistance is zero.

1.3.5 The Conclusions from Section 1.3

The internal parameters of the transmitting antenna include the active component of its input resistance, R_A, equal to the sum of the radiation resistance R_R and the loss resistance R_L, the reactive component of the input resistance X_A and the efficiency (η). Active resistances are the measures of the active power emitted and lost by the antenna, and reactive resistance is a measure of the reactive power at the antenna input.

To evaluate the antenna input matching, it is necessary to know, in addition to R_R, R_L, and X_A, the value of the wave resistance of the transmission line connecting the antenna to the source (transmitter). The degree of antenna input matching is evaluated by numerical values of one of three parameters: $|\Gamma|$, VSWR, or L_{RET}.

1.4 The External Parameters of Transmitting Antennas

The external parameters of the transmitting antennas describe their ability to concentrate the radiated energy in the desired direction and provide the required phase and polarization properties of the emitted signals. All the external parameters are determined for the far zone of the antenna, in which there are no induction fields, and the angular spatial distribution of the radiated energy is already formed and does not depend on the distance to the antenna.

The external parameters of the transmitting antenna include radiation pattern (amplitude, phase, and power ones) and some characteristics of these patterns, such as the main lobe width, the side lobes level, the back lobe level, etc. [1, 2]. The important external parameters are also the directivity, the gain, and the effective area of the antenna.

The general definition of the radiation pattern (RP) of the transmitting antenna can be the following [2]: RP is the directional distribution in a two-dimensional or three-dimensional space of one of the parameters of the electromagnetic field (see Sect. 1.2.1) excited by the antenna in the far zone. RP can be described analytically and depicted graphically (in the form of two-dimensional flat figures, or as the surface of three-dimensional bodies). Let us first give a few brief definitions relating to antenna radiation patterns [2, 3].

Volumetric (3D) RP is the distribution in directions of a three-dimensional space of one of parameters of the electromagnetic (EM) field excited by the antenna in a far zone and the representation of this distribution in the form of a three-dimensional shape (body).

The axial section of the volumetric RP is the RP in the plane passing through the electrical axis of the antenna.

Amplitude RP is the distribution by the directions in a given plane of the strength values of the EM field excited by the antenna in the far zone at equal distances from the phase center of the antenna.

Phase RP is the distribution in the directions in a given plane of the EM field phase values excited by the antenna in the far zone at the same distance from the selected antenna point.

Power RP is the distribution by the directions in a given plane of the square of the strength value of the EM field excited by the antenna in the far zone at equal distances from the phase center of the antenna.

1.4.1 The Amplitude and Power Radiation Pattern

The strength of the field excited by the source in the far zone can be represented as follows:

$$E(r,\theta,\varphi) = \frac{1}{r}\cdot E_0\cdot f(\theta,\varphi)\cdot \exp(-j(kr+\psi(\theta,\varphi)))$$
$$= \frac{1}{r}\cdot E_0\cdot f(\theta,\varphi)\cdot \exp(-j\chi(\theta,\varphi)); \qquad (1.20)$$

where r is the distance from the source to the observation point;

θ, φ are spherical angular coordinates of the observation point;

E_0 is the amplitude factor;

$f(\theta, \varphi)$ is a function that determines the dependence of the amplitude of the field strength excited by the source in the far zone on the angular position of the observation point;

$k = 2\pi/\lambda$ is the wavenumber;

λ is the wavelength in free space;

$\chi(r, \theta, \varphi) = (kr + \psi(\theta, \varphi))$ is a function that determines the dependence of the phase of the field strength excited by the source in the far zone on the angular position of the observation point and its distance from the source.

If the source of excitation of electromagnetic waves is an antenna, the $f(\theta, \varphi)$ function is the three-dimensional amplitude RP of the antenna, the $f^2(\theta, \varphi)$ function is the three-dimensional power RP of the antenna, the $F(\theta, \varphi) = f(\theta, \varphi)/f(\theta, \varphi)_{max}$ function is the normalized three-dimensional amplitude RP, the $F^2(\theta, \varphi)$ function is the normalized three-dimensional power radiation pattern, the $\chi(r, \theta, \varphi)$ function is the three-dimensional phase RP.

Amplitude RP and power RP are the axial sections of the corresponding three-dimensional RPs and are analytically defined as functions of one angle with a fixed value for the other angle: $f(\theta, \varphi = \text{const})$, $f(\theta = \text{const}, \varphi)$, $f^2(\theta, \varphi = \text{const})$, $f^2(\theta = \text{const}, \varphi)$, etc.

Graphical representations of amplitude or power radiation pattern are depicted in Cartesian or polar coordinates. In these graphs, the angle is the independent variable, and the dependent variable is either the amplitude of the radiated field or its power flux density. Dependent variables are expressed in relative units by field or power and normalized to the maximum value (in this case, the maximum value of the dependent variable will be equal to one). Normalized dependent variables are often evaluated in logarithmic measures of the ratio, dB (in this case, the maximum value of the dependent variable will be 0 dB).

These days, the graphical representation of volumetric RP is easily produced with application packages (e.g., MATLAB).

Figures 1.4, 1.5 and 1.6 show examples of graphical representation of normalized amplitude RP and power RP in Cartesian coordinates (Fig. 1.4 is for the amplitude, Fig. 1.5 is for power, and Fig. 1.6 shows the value in dB).

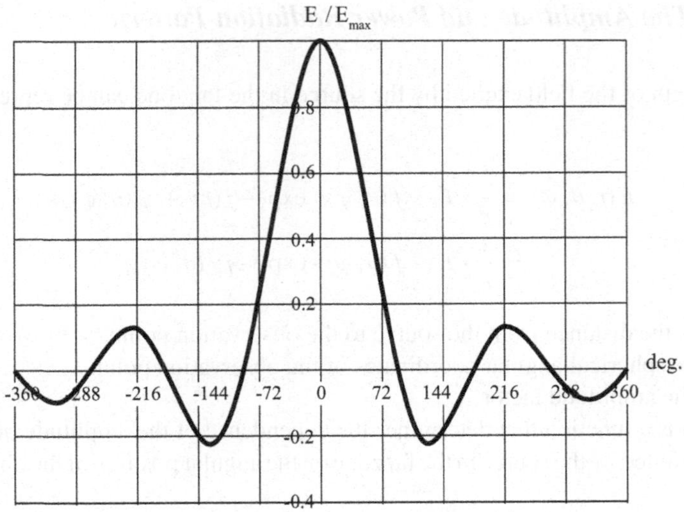

Fig. 1.4 Normalized antenna amplitude PR in Cartesian coordinates

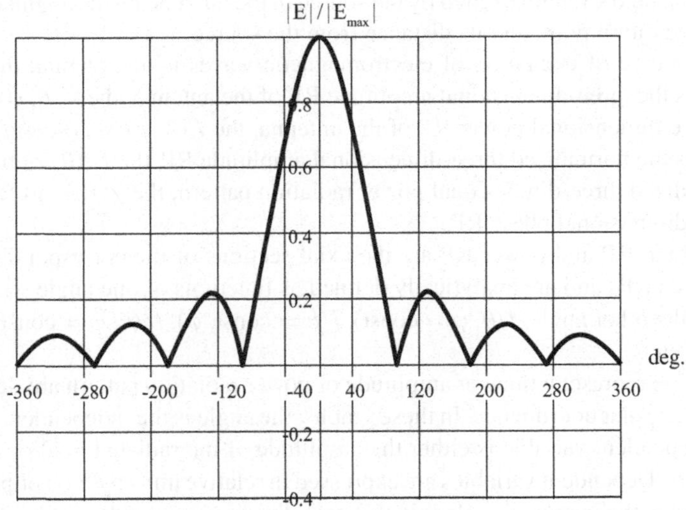

Fig. 1.5 Normalized antenna power RP in Cartesian coordinates

Figure 1.7 shows normalized antenna RP for power in polar coordinates. Figure 1.8 shows an example of a graphical representation of a 3D RP.

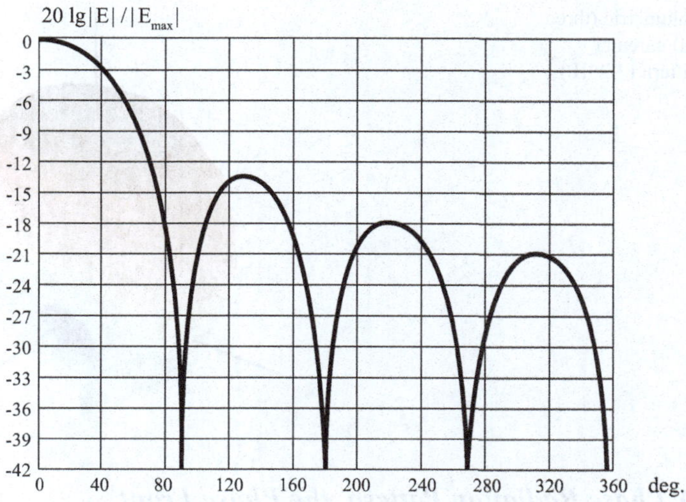

Fig. 1.6 Normalized antenna RP in logarithmic units

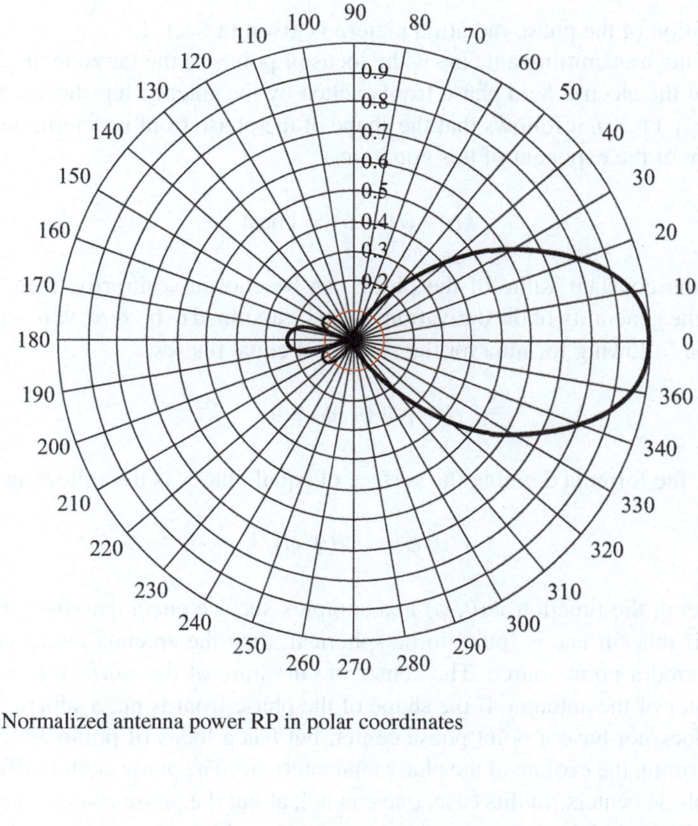

Fig. 1.7 Normalized antenna power RP in polar coordinates

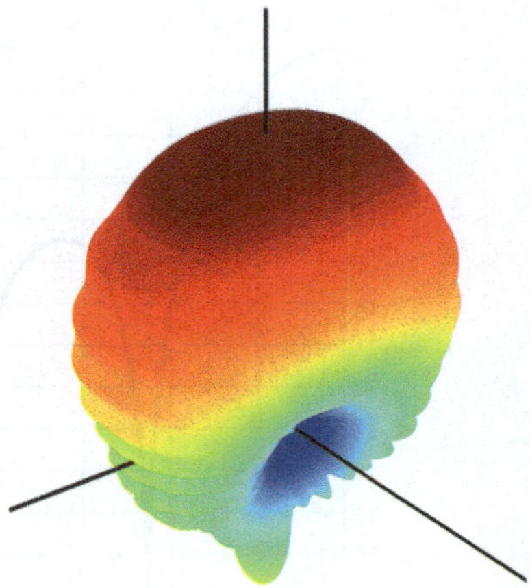

Fig. 1.8 Volumetric (three dimensional) antenna radiation pattern (3D RP)

1.4.2 A Phase Radiation Pattern, the Phase Front and the Phase Center of the Antenna

The definition of the phase radiation pattern is given in Sect. 1.4.

The of the transmitting antenna is the locus of points of the far zone in which the strength of the electric field phase front excited by the antenna has the same phase.

From Eq. (1.20), it follows that the shape of the phase front is determined by the superscript of the exponent of this equation.

$$kr + \psi(\theta, \phi) = \text{const} \tag{1.21}$$

Since the constant value of the phase can be chosen arbitrarily, then, without violating the generality of the derivation, it can be assumed to be zero, which provides us with the following formula for the surface of equal phases.

$$kr + \psi(\theta, \phi) = 0 \tag{1.22}$$

Hence, the formula defining the surface of equal phases is the following:

$$r(\theta, \phi) = \psi(\theta, \phi)/k \tag{1.23}$$

In general, the function $\psi(\theta, \varphi)$ is a complex surface encompassing a radiating antenna. If this surface is found to be spherical, then the antenna under study can be considered a point source. The center of curvature of this surface is called the phase center of the antenna. If the shape of the phase front is not a sphere, then the antenna does not have a point phase center, but has a locus of points of curvature centers forming the evolute of the phase characteristic. The phase center diffuses into a line of phase centers. In this case, one can talk about the phase center for a limited space angle, in which electromagnetic waves excited by the antenna propagate or about the integral phase center [4, 5]. The knowledge of the phase characteristics is particularly important when the antenna is intended for use as an exciter of reflector antenna or as an element of a phased array antenna.

1.4.3 Polarization Characteristic

The polarization plane of the field radiated by the antenna is the plane in which the electric field intensity vector $\boldsymbol{E}$ and the power flux density vector $\boldsymbol{\Pi}$ (Poynting vector) are located.

If the position of this plane in space does not change over time, then the antenna is linearly polarized.

If the polarization plane of the field radiated by the antenna rotates over time around the propagation direction, the antenna has a rotatory polarization with the rotation in the left or the right direction.

If in the rotation of the polarization plane the amplitude of vector $\boldsymbol{E}$ does not depend on the angle of rotation of the plane, the rotatory polarization is called circular, and if there is the dependence, the polarization is elliptical.

The end of the $\boldsymbol{E}$ vector of a wave with rotating polarization for a period of time equal to the period of oscillations draws a closed curve. The projection of this curve on a plane perpendicular to the direction of propagation of the wave provides a visual representation of the polarization type. For a wave with rotatory circular polarization, this curve is a circle. For an elliptically polarized wave, this curve is an ellipse called a polarization ellipse. The ratio of the minor semi-axis of the polarization ellipse to the major one is called the ellipticity or polarization factor [2, 3]. Linear and circular polarizations can be considered special cases of elliptic polarization: For linear one, this is the case when the polarization factor is zero, for the circular one, when the polarization factor is unity.

The polarization characteristic of the antenna is the dependence of the polarization factor of the radiated field (in the far zone) on the direction to the observation point, provided that the distance to the observation point remains constant. Sometimes, when determining the polarization characteristic, the slope of the major semi-axis of the polarization ellipse relative to a given direction is also taken into account.

1.4.4 The Directivity and Gain

The directional transmitting antenna produces a high concentration of radiated energy in a given direction. The advantage in the value of the radiated power density obtained in the direction of the maximum radiation of the directional transmitting antenna compared to a perfectly matched reference antenna without losses is accessed by two parameters, the directivity (D) and the gain (G).

The directivity shows the value of the above-mentioned advantage when both compared antennas emit the same total power, and the gain, when the same power is accepted by the inputs of the antennas to be compared (see Fig. 1.2). Thus, the directivity access the gain of an antenna without taking its active losses into account, while gain access the gain of the antenna with due account for its active losses. The measurement units of directivity and gain are relative units (in terms of power) or dB. There is an obvious relationship between directivity and gain in relative units and in dB:

$$D[\text{dB}] = 10\ \lg\ D\,[\text{relative units}].$$

$$G[\text{dB}] = 10\ \lg\ G\,[\text{relative units}].$$

A reference antenna used in determining the directivity is, as a rule, an isotropic radiator, i.e., a hypothetical antenna that creates the same flux density of radiated power in all directions. However, in some cases, a half-wave vibrator (dipole) is used as a reference antenna when determining the directivity. It is obvious that between the directivity of the antenna defined relative to the isotropic radiator and the directivity of the same antenna defined relative to the dipole, there is a difference equal to the directivity of the dipole itself, which is defined relative to the isotropic radiator.

The directivity of an isotropic radiator is equal to one (times in terms of power), or 0 dB. The directivity of the dipole relative to the isotropic radiator is equal to 1.64 relative units (1.64 times in terms of power) or 2.15 dB.

In order to avoid confusion, an antenna directivity in logarithmic units relative to the isotropic radiator is denoted with dBi, and relative to the dipole, with dBd. There is an obvious relation between the directivity of the same antenna in dBi and dBd:

$$D[\text{dBd}] = D[\text{dBi}] - 2.15. \tag{1.24}$$

Everything stated above with respect to the units of the directivity is true for the units of the gain. By default, the measurement units of directivity and gain in dB correspond to dBi.

There is the following simple relation between directivity and gain in relative units:

$$G = D \cdot \eta. \tag{1.25}$$

If G, D and η are expressed in dB, then the relationship between them is as follows:

$$G[\text{dB}] = D[\text{dB}] + \eta[\text{dB}]. \tag{1.26}$$

(it should be remembered that the η expressed in dB is a negative value).

Sometimes, the angular dependence of the directivity and gain of an antenna is defined as the ratio of the power flux density radiated by the antenna into an infinitely small space angle, the axis of which coincides with the direction under consideration, to the power flux density radiated by an isotropic radiator into the same angle.

When determining the gain of a real antenna, it is necessary to distinguish between the maximum gain with an ideal matching of the antenna input and the realized gain, in the determination of which the mismatch of the antenna input is taken into account. If we turn to Fig. 1.2, then in case of a mismatch of the antenna input it receives not all the power of the P_{TRM} but only its part equal to $P_{TRM}(1\text{-}|\Gamma|^2)$. At the same time, the isotropic antenna, which the antenna under consideration is compared with in the determination of the gain, has a perfectly matched input and receives all the P_{TRM} power. To take this specificity into account, the parameter of realized gain (G_R) is introduced. There is an obvious connection between G and G_R:

$$G_R = G(1 - |\Gamma|^2). \tag{1.27}$$

Gain measurement methods are discussed in Chap. 3 of this book.

1.5 Radio Characteristics of Receiving Antennas

1.5.1 The Identity of Some Radio Characteristics of Antennas in the Transmission and Reception Modes

One of the fundamental statements of the circuit theory is the reciprocity principle, which establishes for linear medium a cross-connection between the sources and the fields created by them at the locations of these sources. [1, 4]. In relation to antennas (which do not include non-reciprocal elements), the reciprocity principle allows one to prove that the antenna radiation patterns are the same in the transmission and reception modes [1, 4].

This leads to an important conclusion: The external parameters of the antenna under study can be measured both when the antenna is receiving and when it is transmitting. The determining criterion for the selection of a measurement method is the simplicity of the measuring facility and the convenience of measurements (these are the only characteristics that matter).

However, the coincidence of the shape of the angular dependence of the external parameters of the antenna when operating in the transmission and reception modes does not mean the same physical nature of these parameters.

In the transmission mode, RP is characterized by the angular distribution of the amplitude, phase or power of the electromagnetic wave radiated by the antenna in the far zone.

In the reception mode, the amplitude, phase, or power of the current in the matched load at the output of the receiving antenna depend on the direction of the plane electromagnetic wave incident on the antenna [1].

1.5.2 The Directivity, Gain, and the Effective Area of the Receiving Antenna

The numerical values of gain, the efficiency, and directivity of the antenna in the reception mode and in the transmission mode coincide. The physical meaning of these parameters for the receiving antenna can be defined as follows.

The gain of the reception antenna is defined as the ratio of the power transferred by the studied antenna into a matched load to the power transferred to the matched load by the isotropic antenna, provided the power flux density of plane electromagnetic waves incident on the antenna is the same.

The efficiency of the receiving antenna shows how many times the power transmitted to the matched load by the antenna under study is lower than the power that this antenna could transmit to the matched load in the absence of active losses.

The directivity of reception antenna is defined as the ratio of the power transferred by the antenna under examination, in the absence of an active loss, toward the matched load, to the power transferred to the matched load by an isotropic antenna, provided the power flux density of plane electromagnetic waves incident on the antenna is the same.

A specific parameter of aperture receiving antennas is the effective area (A_{EFF}) of the antenna, which is understood as the geometric aperture area of a hypothetical perfectly matched antenna which has a directivity equal to the directivity of the real antenna but receives and transmits to the output all the power incident on it. The effective area of an antenna can also be defined as the ratio of the power transferred by the receiving antenna to the matched load to the power per unit area in the plane wave incident on the antenna.

In antenna theory, it is proved that the directivity and A_{EFF} of a perfectly matched lossless antenna ($|\Gamma|^2 = 0$, $\eta = 1$) have the following relation [1, 4]:

$$A_{\mathrm{EFF}} = \frac{\lambda^2}{4\pi} D. \tag{1.28}$$

If the antenna is perfectly matched but has active losses ($|\Gamma|^2 = 0$, $\eta < 1$), then with the same incident power, a lower power is outputted, and the $\mathrm{A_{EFF}}$ of such an antenna will be determined not on the basis of the directivity but on the basis of gain:

$$A_{\mathrm{EFF}} = \frac{\lambda^2}{4\pi} \cdot G = \frac{\lambda^2}{4\pi} \cdot D \cdot \eta. \tag{1.29}$$

Finally, if the antenna is mismatched and has active losses, ($|\Gamma|^2 > 0$, $\eta < 1$), then with the same incident power, it is able to transfer even less power to the receiving path, and the A_{EFF} of such an antenna is determined through the G_{R}:

$$A_{\mathrm{EFF}} = \frac{\lambda^2}{4\pi} \cdot G_R = \frac{\lambda^2}{4\pi} \cdot G \cdot (1 - |\Gamma|^2) = \frac{\lambda^2}{4\pi} \cdot D \cdot \eta \cdot (1 - |\Gamma|^2). \tag{1.30}$$

1.5.3 *The Noise Temperature of the Antenna*

Even in the absence of a useful signal or a source of artificial radio interference, some active power called antenna noise power is found on the matched load of the receiving antenna. The source of these noises is the thermal radiation of the antenna structural elements, as well as the cosmic background radiation received by the antenna and the radio emission of discrete space sources: the Sun, the Moon, and Stars.

It is known that with any active resistance *R*, at the absolute temperature of this resistance equal to *T*, noise electromotive force (EMF) develops. The spectral density of the mean square of this EMF ($|e_{\omega}^{2}|$) is determined by the Nyquist formula [1]:

$$\left|e_{\omega}^{2}\right| = \frac{2}{\pi} R k_{\mathrm{B}} T \tag{1.31}$$

where k_{B} is the Boltzmann constant.

The spectral noise density of the antenna can be expressed in a similar way if the antenna noise temperature (T_{A}) is introduced, which means the temperature to which the radiation resistance of the antenna under study (R_{R}) should be heated so that it gives the same spectral noise power density ($|e_{\omega}^{2}|_{\mathrm{A}}$) as the antenna.

$$\left|e_{\omega}^{2}\right|_{\mathrm{A}} = \frac{2}{\pi} R_{\mathrm{R}} k_{\mathrm{B}} T_{\mathrm{A}},\ T_{\mathrm{A}} = \frac{\pi \left|e_{\omega}^{2}\right|_{\mathrm{A}}}{2 R_{\mathrm{R}} k_{\mathrm{B}}} \tag{1.32}$$

The concept of noise temperature is widely used in radio astronomy methods for measuring antenna parameters.

1.5.3.1 Antenna Frequency Band

The most general definition of the antenna frequency band for both the reception mode and the transmission mode can be formulated as follows: The frequency band is the value of the frequency interval in which the parameters and characteristics of the antenna do not go beyond the specified limits. Depending on the value of the operating frequency band, antennas can be classified into narrowband (resonant), broadband (non-resonant), and ultra-wideband ones.

References

1. Zhuk MS, Molochkov YuB (1966) Planning of antenna-feeder devises. Energy, Moscow, p 648. (in Russian)
2. Geruni PM (1990) Microwave antennas. Antenna measurements. Terms and definitions. VNIIRI Publishing House, Yerevan, p 128. (in Russian)
3. IEEE Standard Board (1993) IEEE standard definitions of terms for antennas

4. Azenberg GZ, Yampolsky VG, Tereshin ON (1977) VHF antennas, part 1. "Communication" publishing, Moscow, p 382. (in Russian)
5. Azenberg GZ, Yampolsky VG, Tereshin ON (1978) VHF antennas, part 2. "Communication" publishing, Moscow, p 287. (in Russian)

Chapter 2
Methods of Measurement of Radio Characteristics of Antennas (Brief Overview)

2.1 Introduction

Currently, many electrodynamic problems are solved with the use of powerful application packages that make it possible to solve these problems with high accuracy. In particular, calculations of the radio characteristics of different antennas are performed with Ansoft HFSS, Microwave Studio, Microwave Office, and some other application packages. The question may arise whether we need antenna measurements at all. After all, with the same accuracy of the result obtained by measurement and by calculation, the use of application packages has its obvious advantages.

At the same time, there are a number of reasons that make it impractical or impossible to avoid antenna measurements. These include:

1. The limited possibilities of computer models for taking into consideration such subtle matters as:
   - The effect of the carriers on which the antennas are installed on the radio characteristics of these antennas;
   - The influence of the dispersion of parameters of the materials which antennas are made of, on radio characteristics of these antennas;
   - The influence of external dynamic factors (e.g., rain, snow, aerodynamic heating, etc.) on the radio characteristics of the antennas.
2. The need for measurements in connection with the performance of the following works:
   - Certification of reference products;
   - The confirmation of reliability of computer models.

So, having made sure that it is still too early to replace antenna measurements completely with the results of mathematical modeling, let us proceed to the description of the measuring methods for radio characteristics of antennas.

The radio characteristics of the antenna can be measured in near-field and far-field regions.

V. S. Kalashnikov et al., *Near-Field Antenna Measurements*, Springer Aerospace Technology, https://doi.org/10.1007/978-981-33-6436-3_2

The main methods of measurement in the far zone include outdoor far-field range method, aerial method, and radio astronomy one. In far-field methods, the response of the antenna to the plane wave is obtained immediately in the measurement process.

Other methods include the collimator (compact test range) method, near-field methods, and combined method. In the collimator method, the response of the test antenna to the plane wave is also obtained immediately in the measurement process. In the case of the near-field method, the amplitudes and phases of the tangential components of the electric (or magnetic) field strength vector on a given surface near the antenna are determined experimentally, and then the experimental data arrays are used to calculate the radio characteristics of the studied antenna with the use of special formulas. In the combined method, the main and the nearby side lobes of RP of highly directional antennas are determined by the near-field method, while far side lobes are measured at the same facility, but with the far-field method [1, 2].

The object of the research in this monograph is the near-field methods of measurement, and, therefore, the other methods will only be briefly mentioned. We can recommend [1, 2] to find out more about them.

2.2 Measurements in the Far Zone

2.2.1 Outdoor Far-Field Range Method

When the AUT operates in the transmission mode, the radiation patterns are created on the basis of the readings of a field indicator which is fixed in the far zone of the AUT, depending on the rotation angle of the AUT in the plane under examination. When the AUT is operating in the receiving mode, the radiation pattern is created on the basis of the measured power received by the AUT. In this case, the AUT should rotate in the plane under examination to any angle if the auxiliary source of electromagnetic radiation is stationary, or remain stationary when the auxiliary source is moving.

The method of measuring radiation patterns using the radiation field of the AUT or an auxiliary antenna located in the far zone of the AUT is traditional and the most widely used. The AUT and the auxiliary antenna are located at a certain distance from each other at altitudes that provide direct visibility and the absence of interfering objects in the communication line. The mutual position of the antennas is set in the way that reduces the impact of ground echo. In this case, either the auxiliary antenna or the AUT, or both antennas are located on masts or towers. In their turn, masts or towers are placed on a special testing range, where the most favorable conditions are created to reduce measurement errors. This method has a number of advantages. First, the auxiliary antenna (which is usually a transmitting antenna) is located on the tower and can be fixed to remain motionless. This makes it possible to sight the AUT on the radiator (to ensure the alignment of the AUT) by quite simple methods. Secondly, a generator installed next to or near the auxiliary antenna can emit a signal of any shape and level. Third, the measurement conditions, except the weather, hardly change over time. Fourthly, the received signals do not require special processing, because they are the direct response of the receiving antenna to the incident plane electromagnetic wave. All this ensures the ease of organization and performance of measurement and ensures the stability and reliability of the results.

However, the method described here also has significant drawbacks. First, the far-field range method does not allow to measure the three-dimensional radiation pattern of the AUT. Secondly, there are great difficulties involved in the reduction of the impact that electromagnetic waves reflected from the ground and the local objects have on the measurement results. Third, for large antennas with narrow radiation patterns (antennas with large electrical sizes of the aperture), the lower boundary of the far zone can be hundreds of meters or even several kilometers, and it becomes impossible to ensure the required distances between the AUT and the auxiliary antenna. Fourth, the outdoor far-field range method cannot be used to tune phased antenna arrays or active phased antenna arrays.

More detailed information about the outdoor far-field range method for antenna radio characteristic measuring can be found in [1, 2].

2.2.2 The Aerial Method

In the case of large size ground-based fixed antennas to measure RP and the potential, an auxiliary antenna has to be installed on an aircraft, which can be an airplane, a helicopter, or an unmanned aerial vehicle (drone) [3–5]. The specificity of the method requires the introduction of the maximum level of automation for all stages of the work, from flight control, measurements, and registration of the results, to information processing and the issue of documents with the characteristics of the AUT. The aerial method is also used to measure antennas of another category: weakly directional on-board antennas. In this version, the aerial method is much easier than when it is necessary to fly around large antennas. However, there are difficulties associated with determining the coordinates of the aircraft and the recording of the measurement results. Unlike all other measurement methods, the aerial method deals with a mobile auxiliary antenna moving in space along a trajectory that can be specified only in a statistical sense, i.e., in a certain range of possible values relative to the required ones. Therefore, in this method, the problem of determining the position of the auxiliary antenna (or the AUT, in the case of on-board antennas) becomes complex and extremely necessary. It is required to determine all the three coordinates: both angular ones and the slant range.

More detailed information about the aerial method of measuring the radio characteristics of antennas is provided in [1].

2.2.3 *The Radio Astronomy Method*

The field of application of radio astronomy methods of RP measurement is large antennas and antenna arrays of radio telescopes. In the case of using the radio astronomy method, the AUT operates in reception mode, and the irradiating field is created by extraterrestrial sources of noise radio emission, e.g., the moon, stars, nebulae, etc. These methods have numerous limitations: RP can be measured only in one plane, the dynamic range of the measurements is low, which makes it impossible to reliably determine the level of the side lobes, the frequency characteristics of the RP cannot be measured, etc.; however, for some types of the above-mentioned antennas experimental determination of radio characteristics by other methods is simply impossible. More detailed information about the radio astronomy method of measuring the radio characteristics of antennas can be found in [1].

2.3 The Compact Antenna Test Range Method (CATR, the Collimator Method)

The collimator method consists in irradiating the examined antenna with an electromagnetic wave with a plane phase front created by a special auxiliary antenna referred to as a collimator. As a rule, a collimator is a reflector antenna consisting of a parabolic reflector and an exciter. An example of a collimator reflector is shown in Fig. 2.1.

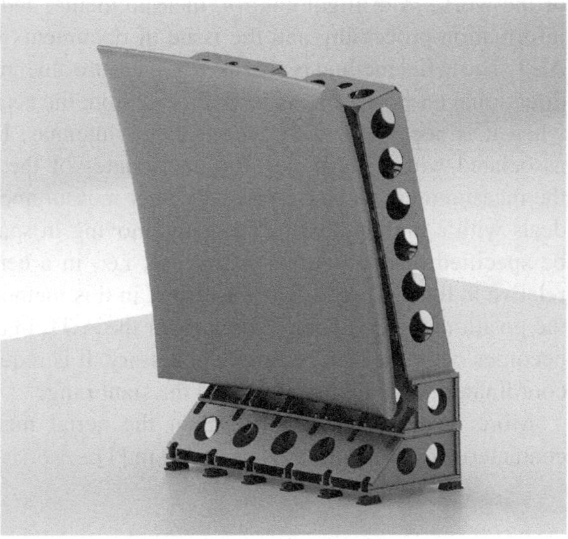

Fig. 2.1 A collimator reflector with rounded edges (Smitek company)

The reflector converts the divergent spherical wavefront of the wave created by the exciter into a plane wavefront. The collimator field, which is close to the plane wave field, is formed in a limited area of space located in front of the aperture (see Sect. 2.2). The main reasons for the deviation from the constant values of the amplitude and the phase of the field in this region are the diffraction effects caused by the limited size of the reflector and the unevenness of the field, which is due to the directivity of the exciter.

The measuring facilities with the use of collimators are called compact antenna test range (CATR).

Some examples of compact antenna test ranges are shown in Figs. 2.2 and 2.3. There are three different implementations of such test ranges [6].

Single-reflector compact range

It has a very simple arrangement and uses only one auxiliary parabolic reflector to convert the spherical wave of the exciter into a wave with a plane phase front within a limited area of space near the aperture of the reflector. This area is called the test zone of the collimator and is determined by its transverse dimensions and depth. The derivation of formulas for determining the size of the test zone of a single-reflector collimator is given in [2].

Single-reflector collimators of two configurations were developed: a collimator with a linear source, consisting of a reflector in the form of a parabolic cylinder with

Fig. 2.2 A compact antenna test range (Smitek company)

Fig. 2.3 Compact antenna test range, a mobile version (Smitek company)

a large hoghorn exciter, and a collimator with a point source consisting of a large parabolic reflector and a small horn exciter. A collimator with a linear source makes it possible to obtain a purely linear polarization thanks to a cylindrical reflector. The disadvantage of this collimator is that when the frequency is changed, it is necessary to change the physical size of the aperture of the linear horn exciter in order to preserve the required field distribution in the aperture of the parabolic cylinder. The collimator with a point source is more advanced. The mirror of such a collimator can operate in a wide frequency band.

Double-reflector compact range

Unlike a single-reflector compact test range, this test range, in addition to the main reflector in the form of a parabolic cylinder, which adjusts the wavefront in one plane, has an additional subreflector to adjust the wavefront in another plane. When their dimensions are comparable, a double-reflector test range provides a much greater depth of the test zone than a single-reflector compact antenna test range.

Compensated compact antenna test range

This testing range also has two reflectors. However, due to the use of special-shaped reflectors (with double curvature for hyperbolic and parabolic functions), the emergence of the intrasystem cross-polarization is largely prevented. This means that there is a good uncoupling between the components of the electromagnetic field with orthogonal polarizations.

2.4 The Near-Field Measurement Method

As mentioned above, the implementation of the near-field method involves experimentally determining the amplitudes and phases of the tangential components of the electric (or magnetic) field strength vector on a given surface near the antenna, and then the experimental data arrays are transformed into the radio characteristics of the AUT with the use of special formulas. Measurements are performed with an auxiliary antenna acting as a probe and a special device called ampliphasemeter; currently, this function is implemented with vector network analyzers. The surface on which the probe is moved and on which the amplitude and phase of the tangential component of the field strength vector are measured is called the scan surface. Ideally, the scan surface should encompass the antenna under investigation. In practice, three forms of scan surfaces are considered preferable compared to the others; these are the plane, the side surface of a circular cylinder, and the sphere. The openness of the first two is compensated by their sizes (theoretically, they are infinite, but in practice, large enough to provide amplitude at the edge lower than the maximum one by 30–40 dB). The preference given to these forms of scanning surfaces is explained by the relative simplicity of kinematic schemes of corresponding measuring facilities and the relative simplicity of computational operations that use the measured values of the modulus and the phase of the tangential components of the electric field vector on the scanning surface to calculate the radio characteristics of the AUT [7].

2.5 The Combined Method

In some cases of practical importance, there is a need for experimental evaluation of the level of the far side lobes of highly directive antennas, which can be equal to—(50–70) dB. At the same time, the errors of determination of the levels of such RP with the use of near-field methods are quite sizeable. To solve the above problem, a combined method for determining the RP of antennas in the near zone was developed.

The combined method is based on the fact that the structure of the lateral radiation of highly directive antennas peculiar to the field in the far zone (in particular, the law of the change of the far side lobe envelope) is formed at a distance much smaller than $2D^2/\lambda$. In other words, near the aperture of a highly directive antenna, there is a region of angles in which the distribution of the field, up to a slowly changing multiplier, coincides with the distribution in the far zone. In this area, it is possible, for the purpose of RP determination, to replace near-field to far-field transformation by near-field normalization, virtually eliminating the methodological and instrumental errors inherent in the near-field method used to calculate the RP for low levels of the measured field (Fig. 2.4). Due to this, the overall measurement error is reduced and the dynamic range of the determined RP levels of the AUT is expanded, reaching (70–90) dB, and is limited mainly by the influence of reflections from surrounding

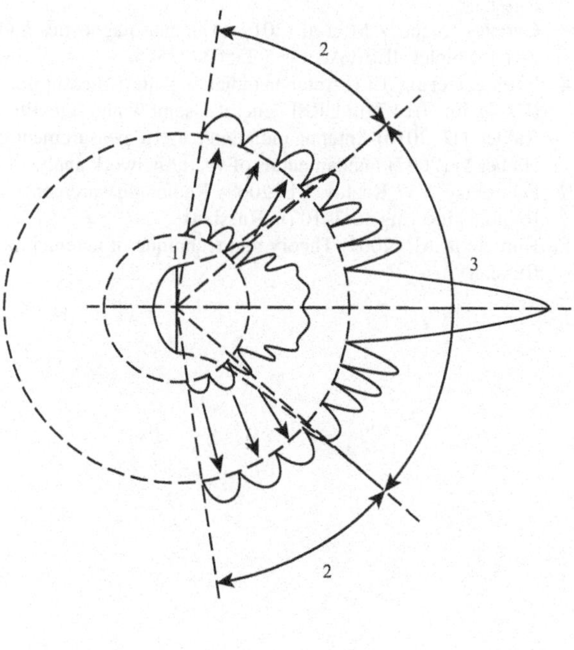

Fig. 2.4 Combined method. 1—antenna, 2—the area of the RP calculation using normalization, 3—the area of the RP calculation using near-field to far-field transform

objects, which can be made very insignificant in case of measurements in anechoic chamber.

In order to implement the combined method, it is necessary to solve the following basic tasks:

- Determine the area of angles at a known distance from the antenna to the scan surface, for which the envelopes of the side lobes in the far and near zones are similar (area 2 in Fig. 2.4);
- To find the normalization factor (similarity coefficient), which is equal to the ratio of the level of the envelope of the lateral radiation (or some of its functions) directly measured in the near zone and the level of the RP, calculated with the near-field transform.

Thus, when the combined method is used, the results of the measurements of the amplitude–phase field distribution near the AUT undergo the following processing: In the space angle covering the main and near side lobes, the RP is determined by mathematical transformations used in the near-field method, and for the region of the far side lobes having a low level, by normalizing the measurement results [1, 2, 8].

References

1. Zakharyev LN, Lemansky AL, Turchin VI, Tseytlin NM et al (eds) (1985) Methods for measuring the characteristics of microwave antennas. Radio Commun Moscow 368. (in Russian)
2. Bahrakh LD, Kurochkin AP, Kremenetsky SD, Usin VA, Shifrin YS (1985) Methods for measuring the parameters of radiating systems in the near-field. Leningrad, Nauka, 272 p. (in Russian)
3. García-Fernández M et al (2017) Antenna diagnostics and characterization using unmanned aerial vehicles. IEEE Access 5:23563–23575
4. Üstüner F et al (2014) Antenna radiation pattern measurement using an unmanned aerial vehicle (UAV). In: 2014 31th URSI general assembly and scientific symposium (URSI GASS). IEEE
5. Ridder TD (2016) Antenna radiation pattern measurement using an unmanned aerial vehicle
6. Hiebel M (2007) Fundamentals of vector network analysis. Rohde & Schwarz
7. Denisenko VV, Kozlov YI (2008) Radio measurements in specialized anechoic chambers. Radioengineering 10:3–10 (in Russian)
8. Kurochkin AP (2009) Theory and technique of antenna measurements. Antennas 7:39–44 (in Russian)

Chapter 3
Planar Near-Field Antenna Measurements: Calculation Expressions

3.1 Electrodynamic Fundamentals of the Amplitude–Phase Measurements

In theoretical electrodynamics, there is a rigorous solution to the problem of determining the electromagnetic field excited by sources (electric and magnetic currents and charges) at an arbitrary point P inside the volume V, bounded from the inside by closed surfaces $S_1 \ldots S_{n-1}$, and from the outside by a closed surface S_n (Fig. 3.1).

This solution is the sum of the volume integral over the sources located inside V and the surface integrals over the electric and magnetic fields on the surfaces $S_1 \ldots S_n$. The latter are assumed to be excited by sources located in closed surfaces $S_1 \ldots S_{n-1}$ and outside the surface S_n [1–3]. For the complex amplitude of the vector $\mathbf{E}$ at the point P, this solution is as follows:

$$\mathbf{E_P} = \frac{1}{4\pi}\int\limits_V \left\{\mathrm{j}\omega\mu\mathbf{J}\psi + \left[\mathbf{J_m}, \text{grad}\ \psi\right] - \frac{\rho}{\varepsilon}\text{grad}\ \psi\right\}\mathrm{d}V$$
$$+ \frac{1}{4\pi}\int\limits_S \left\{-\mathrm{j}\omega\mu\psi[\mathbf{n}, \mathbf{H_s}] + \left[[\mathbf{n}, \mathbf{E_s}], \text{grad}\ \psi\right] + (\mathbf{n}, \mathbf{E_s})\text{grad}\ \psi\right\}\mathrm{d}S, \quad (3.1)$$

where $\mathbf{J}$ is the conduction current density; $\mathbf{J_m}$ is the magnetic current density; ρ is the electric charge density; ρ_m is the magnetic charge density; $\mathbf{E_s}$ is the electric field strength on the surfaces $S_1 \ldots S_n$; $\mathbf{H_s}$ is the magnetic field strength on the surfaces $S_1 \ldots S_n$; r are the distances from integration points to the point P at which the field is calculated;

$\psi = \exp(-jkr)/r$ is the Green function; $k = 2\pi/\lambda$ is the wave number; and $\mathbf{n}$ is the normal to the surface S on which fields $\mathbf{E_s}$ and $\mathbf{H_s}$ are defined.

If all sources of the field are concentrated inside the closed surface S, and the sighting point P is outside this surface (see Fig. 3.2), then, depending on what is known about these sources, the field at point P can be found in two ways:

V. S. Kalashnikov et al., *Near-Field Antenna Measurements*, Springer Aerospace Technology, https://doi.org/10.1007/978-981-33-6436-3_3

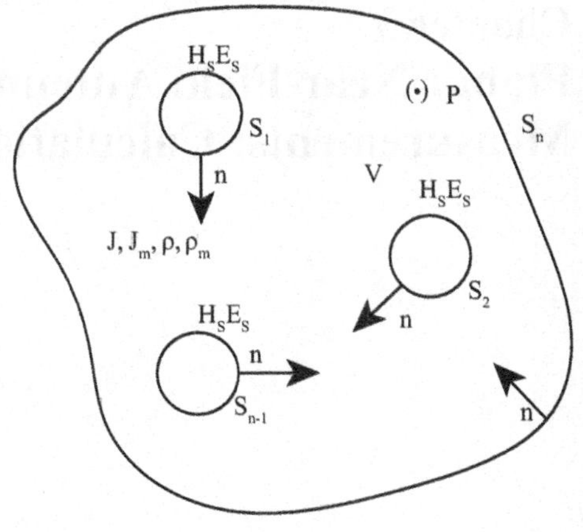

Fig. 3.1 Finding field inside volume V

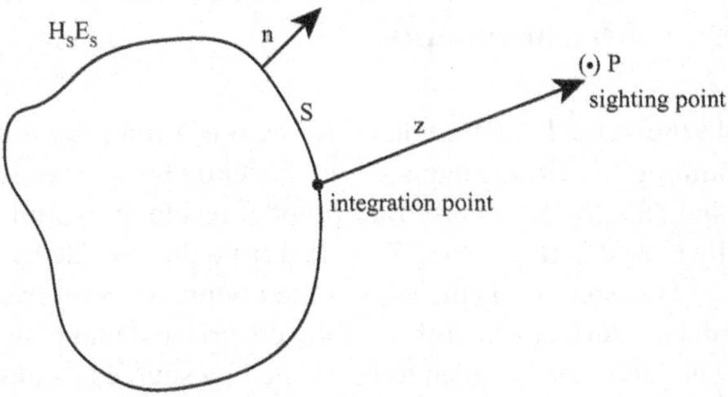

Fig. 3.2 Calculating the field outside of the closed surface S

1. If the distribution of sources (J, J_{m}, ρ, ρ_{m}) inside the volume V bounded by the surface S is known, then the field at the point P is determined by taking the volume integral over these sources, and the surface integral vanishes. The expression for the field at point P in this case has the following form:

$$\mathbf{E}_{\mathbf{P}} = \frac{1}{4\pi} \int\limits_{V} \left\{ \mathrm{j}\omega\mu\mathbf{J}\psi + \left[\mathbf{J}_{\mathbf{m}}, \operatorname{grad}\psi\right] - \frac{\rho}{\varepsilon}\operatorname{grad}\psi \right\} \mathrm{d}V \tag{3.2}$$

2. If the values of the electric and magnetic fields on the surface S are known (which are excited by sources located inside this surface), then the field at the point P is determined by taking the surface integral from these fields, and the volume integral vanishes. The expression for the field at point P in this case is the following:

$$\mathbf{E}_{\mathbf{P}} = \frac{1}{4\pi} \int\limits_{S} \left\{ -\mathrm{j}\omega\mu\psi[\mathbf{n}, \mathbf{H}_{\mathbf{s}}] + \left[[\mathbf{n}, \mathbf{E}_{\mathbf{s}}], \operatorname{grad}\psi\right] + (\mathbf{n}, \mathbf{E}_{\mathbf{s}})\operatorname{grad}\psi \right\} \mathrm{d}S. \tag{3.3}$$

Let us introduce special designation for tangential and normal components of surface fields $\mathbf{E_s}$ and $\mathbf{H_s}$. The tangential components are denoted by $\mathbf{E_\tau H_\tau}$, and normal ones—$\mathbf{E_n H_n}$, where:

$$\mathbf{E_\tau} = [\mathbf{n}, \mathbf{E_s}], \qquad \mathbf{H_\tau} = [\mathbf{n}, \mathbf{H_s}],$$
$$\mathbf{E_n} = (\mathbf{n}, \mathbf{E_s}), \qquad \mathbf{H_n} = (\mathbf{n}, \mathbf{H_s}).$$

Then, expression (3.3) takes the following form:

$$\mathbf{E_P} = \frac{1}{4\pi} \int_S \{-j\omega\mu\psi\mathbf{H_\tau} + [\mathbf{E_\tau}, \mathrm{grad}\,\psi] + \mathbf{E_n}\,\mathrm{grad}\,\psi\}dS. \tag{3.4}$$

Expression (3.4) states the fundamental possibility of calculating the field excited by the sources at any point outside the closed surface covering the sources from the known values of the tangential and normal components of the electric and magnetic fields excited by the sources on this surface. As applied to the theory of antennas, this possibility was transformed into the development of methods for determining the antenna far field, based on the solutions of the boundary value problems of electrodynamics for homogeneous wave equations satisfying the radiation conditions at infinity and the boundary conditions on the surface S. Moreover, it is proved that these boundary value problems have a unique solution if any two of the six field components $\mathbf{E_s}$ and $\mathbf{H_s}$ are known on the surface S (six possible components are composed of two orthogonal components of the electric field $\mathbf{E_\tau}$, two orthogonal components of the magnetic field $\mathbf{H_\tau}$ and two normal components—$\mathbf{E_n}$ and $\mathbf{H_n}$) [3]. In solving practical problems, normal components $\mathbf{E_n}$ and $\mathbf{H_n}$ are usually much smaller than tangential ones and can be neglected, and of the remaining four tangential components, the determination of the orthogonal components of the vector $\mathbf{E_\tau}$ is less difficult. Thus, it becomes obvious that if one measures the tangential component complex amplitudes (amplitude and phase) of the electric field vector in set of points on the surface surrounding the antenna, then antenna's RP can be calculated. Moreover, there are no fundamental restrictions on the shape of the surface S and its location relative to the antenna under study. The above considerations underlie the methods for determining antenna far-field characteristics from the measured values of the tangential components of the electric field strength vector on the surface located in the near field of the antenna. This surface is called measurement or scan surface. These methods are called "amplitude–phase methods" or "near-field methods" [3, 4].

In the antenna theory, the problem of determining the amplitude radiation pattern of radiating apertures (whose dimensions exceed the working wavelength) with a uniform phase distribution (the so-called cophased apertures) has been solved. The expressions for calculating the radiation patterns of rectangular and round cophased apertures with various idealized amplitude distributions are given in many monographs on the theory of microwave antennas. In these monographs, the influence of possible differences of the phase distribution from uniform one on the amplitude radiation pattern is considered only for a few idealized cases. These include a linear, quadratic, and cubic phase aperture distribution [1, 2, 5].

The difficulties arising in the development of a mathematical apparatus for amplitude–phase measurements consist in the fact that this apparatus must take into account the phase change voluntary character of the vector **E** tangential component on the scanning surface, as well as the discrete nature of the primary information presented in the form of arrays of experimental data measured at coordinate grid's nodes on the scan surface.

Since the accuracy of the final results of amplitude–phase measurements depends on the step size of this grid, it is necessary to process a large amount of information to increase it. Historically, this problem was first solved by using optical (holographic) processing methods, in which main advantage was efficiency, while the disadvantage was low accuracy, with the following using of special algorithms for digital information processing, namely fast Fourier transform (FFT) and its modifications, which, unfortunately, also did not allow to obtain final results with high accuracy [3, 4].

In recent years, due to the improvement of the PC and the increase in their memory and speed, it has become possible to process a huge amount of digital information in a short time, which allows the use of discrete Fourier transform (DFT) and various methods of interpolation, filtering, and iteration to solve the above problems and achieve the final results, high accuracy.

3.2 Derivation of Calculating Formulas Based on the Kirchhoff Approximation

In the Kirchhoff approximation, the relationship between the electric field vector complex amplitude in the far field of the flat radiating aperture ($\mathbf{E_{FF}}$) and the tangential component complex amplitudes of the electric and magnetic field strength vectors in this aperture ($\mathbf{E}_\tau$ and $\mathbf{H}_\tau$) is defined as follows [3, 4]:

$$\mathbf{E_{FF}} = -\frac{\mathrm{j}}{2\lambda}\frac{\mathrm{e}^{-\mathrm{j}kR}}{R}\int\limits_S [\mathbf{i_R}, \{[\mathbf{n}, \mathbf{E}_\tau] - \mathbf{Z}_0[\mathbf{i_R}, [\mathbf{n}, \mathbf{H}_\tau]]\}]\mathrm{e}^{\mathrm{j}k\rho\cos\gamma}\mathrm{d}S, \tag{3.5}$$

where $k = 2\pi/\lambda$ is the wave number; λ is the wavelength; $\mathbf{R} = R\,\mathbf{i_R}$ is the radius vector of the «current» point in the far-field region; $\mathbf{i_R}$ is the unit vector in the direction of the radius vector **R**; n is the unit normal to the aperture in the direction opposite to the direction of radiation from the aperture; $Z_0 = 120\pi$ is the free space wave impedance; $\boldsymbol{\rho}$ is the radius vector of the "current" point in the aperture; γ is the angle between the radius vectors $\boldsymbol{\rho}$ and **R**; and S is the aperture surface.

For aperture which linear dimensions exceed $\lambda/2$, the tangential components $\mathbf{E}_\tau$ and $\mathbf{H}_\tau$ at each aperture point are connected in the same way as in a plane wave [3, 4]

$$\mathbf{E}_\tau = Z[\mathbf{n}, \mathbf{H}_\tau],\ \mathbf{H}_\tau = (1/Z)[\mathbf{E}_\tau, \mathbf{n}], \tag{3.6}$$

where Z is the so-called aperture wave impedance, which depends on the shape and size of the aperture and on the direction in which the far field is calculated.

In view of (3.6), one can express $\mathbf{E}_{\mathbf{FF}}$ only in terms of the aperture electric field tangential component $\mathbf{E}_\tau$:

$$\mathbf{E}_{\mathbf{FF}} = -\frac{\mathrm{j}}{2\lambda}\frac{\mathrm{e}^{-\mathrm{j}kR}}{R}\int\limits_S [\mathbf{i}_\mathbf{R}, [(\mathbf{n} - (\mathbf{Z}_0/\mathbf{Z})\mathbf{i}_\mathbf{R}), \mathbf{E}_\tau]]\mathrm{e}^{\mathrm{j}k\rho\cos\gamma}\mathrm{d}S. \tag{3.7}$$

For plane apertures which linear dimensions exceed (3…5) λ, one can take $Z = Z_0$ and assume that $\mathbf{n}$ is constant throughout all aperture. In this case, expression (3.7) takes the following form:

$$\mathbf{E}_{\mathbf{FF}} = -\frac{\mathrm{j}}{2\lambda}\frac{\mathrm{e}^{-\mathrm{j}kR}}{R}\int\limits_S [\mathbf{i}_\mathbf{R}, [(\mathbf{n} - \mathbf{i}_\mathbf{R}), \mathbf{E}_\tau]]\mathrm{e}^{\mathrm{j}k\rho\cos\gamma}\mathrm{d}S. \tag{3.8}$$

To obtain numerical results, it is necessary to rewrite expression (3.8) in some coordinate systems that will be related to the "current" aperture point, the "current" sighting point in far-field region, and each other.

Let us overlap aperture with the *XOY* plane of the Cartesian coordinate system *XYZ* and direct *OZ*-axis toward the half-space into which aperture radiates. Figure 3.3 shows the coordinate system in the aperture plane of the antenna. In this coordinate system, the position of the unit normal $\mathbf{n}$, the radius vector of the "current" aperture point $\boldsymbol{\rho}$, and the tangential component $\mathbf{E}_\tau$ are defined as follows:

$$\mathbf{n} = -\mathbf{i}_\mathbf{z}, \tag{3.9}$$

$$\boldsymbol{\rho} = x\mathbf{i}_\mathbf{x} + y\mathbf{i}_\mathbf{y}, \tag{3.10}$$

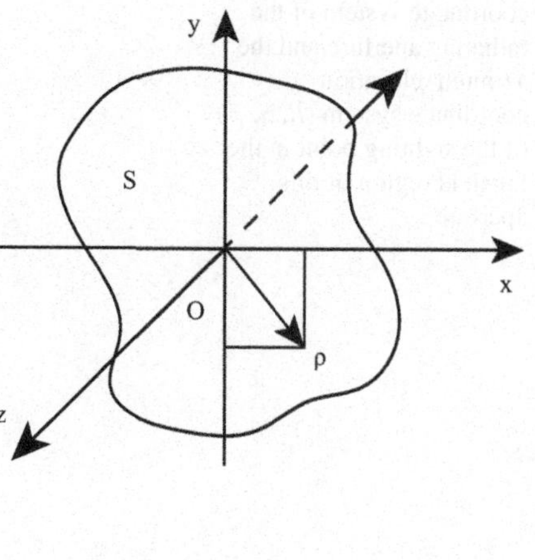

Fig. 3.3 Cartesian coordinate system of the radiating aperture (*XOY*—aperture plane, *n*—normal to the aperture in the negative direction of the *OZ*-axis)

where *x*, *y* are the coordinates of the "current" point in the aperture.

$$\mathbf{E}_\tau = a_x(x, y)\mathbf{i}_\mathbf{x} + a_y(x, y)\mathbf{i}_\mathbf{y}, \tag{3.11}$$

where $a_x(x, y)$, $a_y(x, y)$ are the complex amplitudes of the $\mathbf{E}_\tau$ projections onto the coordinate axes *OX* and *OY*, respectively.

In view of (3.9) and (3.10), expression (3.8) can be rewritten in the following form:

$$\mathbf{E}_{\mathbf{FF}} = \frac{\mathrm{j}}{2\lambda}\frac{\mathrm{e}^{-\mathrm{j}kR}}{R}\int\limits_S \left[\mathbf{i}_\mathbf{R}, \left[(\mathbf{i}_z + \mathbf{i}_\mathbf{R}), \left(a_x\mathbf{i}_\mathbf{x} + a_y\mathbf{i}_\mathbf{y}\right)\right]\right]\mathrm{e}^{\mathrm{j}k\rho\cos\gamma}\mathrm{d}S. \tag{3.12}$$

The position of the "current" sighting point in the far-field region of the radiating aperture will be specified in two coordinate systems combined with *XYZ*—spherical (R, θ, φ) and azimuth–elevation (R, α, β). Figures 3.4 and 3.5 show the coordinate

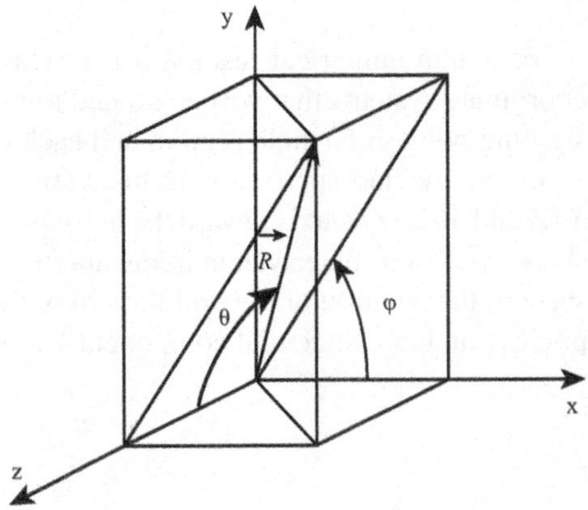

Fig. 3.4 Cartesian coordinate system of the radiating aperture and the spherical coordinate system (R, ϑ, φ) of the sighting point in the far-field region of this aperture

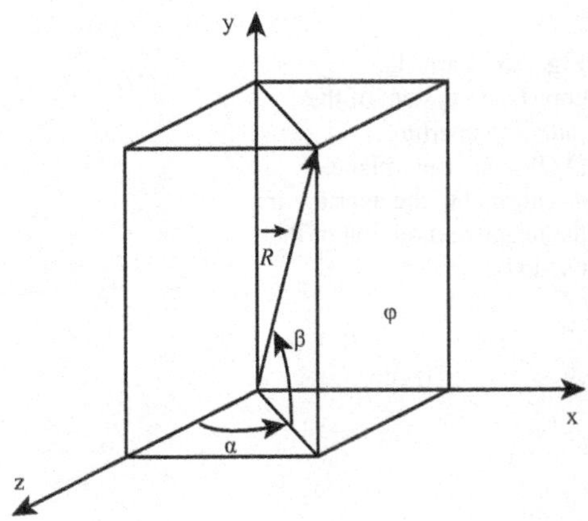

Fig. 3.5 Cartesian coordinate system of the radiating aperture and the azimuth–elevation coordinate system (R, α, β) of the sighting point in the far-field region of this aperture

systems of the radiating aperture and sighting points in the far-field region. For (R, θ, φ) and (x, y, z) coordinate systems, the relationship between the unit vectors $\mathbf{i_R}$, $\mathbf{i_\theta}$, $\mathbf{i_\varphi}$ and $\mathbf{i_x}$, $\mathbf{i_y}$, $\mathbf{i_z}$ is defined by the following coordinate transformation matrix [1]:

$$\begin{pmatrix} \mathbf{i_R} \\ \mathbf{i_\theta} \\ \mathbf{i_\varphi} \end{pmatrix} = \begin{pmatrix} \sin\theta\cos\varphi & \sin\theta\sin\varphi & \cos\theta \\ \cos\theta\cos\varphi & \cos\theta\sin\varphi & -\sin\theta \\ -\sin\varphi & \cos\varphi & 0 \end{pmatrix} \cdot \begin{pmatrix} \mathbf{i_x} \\ \mathbf{i_y} \\ \mathbf{i_z} \end{pmatrix} \tag{3.13}$$

In view of (3.10) and (3.13), one obtains:

$$\rho\cos\gamma = (\mathbf{R}, \boldsymbol{\rho})/R = (\mathbf{i_R}, \boldsymbol{\rho}) = x\sin\theta\cos\varphi + y\sin\theta\sin\varphi \tag{3.14}$$

Since $\mathbf{i_R} = [\mathbf{i_\theta}, \mathbf{i_\varphi}]$, $(\mathbf{i_\theta}, \mathbf{i_x}) = \cos\theta\cos\varphi$, $(\mathbf{i_\varphi}, \mathbf{i_x}) = -\sin\varphi$, $(\mathbf{i_\theta}, \mathbf{i_y}) = \cos\theta\sin\varphi$, $(\mathbf{i_\varphi}, \mathbf{i_y}) = \cos\varphi$, as well as the fact that for large size planar cophase apertures the normal $\mathbf{n}$ can be considered constant throughout all aperture, vector transformations can be removed from the integral, and expression (3.12) takes the form:

$$\mathbf{E_{FF}} = \frac{\mathrm{j}}{\lambda}\frac{\mathrm{e}^{-\mathrm{j}kR}}{R}\left\{\mathbf{e_x}(\theta,\varphi)N_x(\theta,\varphi) + \mathbf{e_y}(\theta,\varphi)N_y(\theta,\varphi)\right\}, \tag{3.15}$$

where

$$\mathbf{e_x}(\theta,\varphi) = \frac{(1+\cos\theta)}{2}\left(\mathbf{i}_\theta\cos\varphi - \mathbf{i}_\varphi\sin\varphi\right), \tag{3.16}$$

$$e_y(\theta,\varphi) = \frac{(1+\cos\theta)}{2}\left(i_\theta\sin\varphi + i_\varphi\cos\varphi\right), \tag{3.17}$$

$$N_x(\theta,\varphi) = \int_S a_x(x,y)\cdot \mathrm{e}^{ik(x\sin\theta\cos\varphi + y\sin\theta\sin\varphi)}\mathrm{d}x\mathrm{d}y, \tag{3.18}$$

$$N_y(\theta,\varphi) = \int_S a_y(x,y)\cdot \mathrm{e}^{ik(x\sin\theta\cos\varphi + y\sin\theta\sin\varphi)}\mathrm{d}x\mathrm{d}y. \tag{3.19}$$

In view of (3.16)–(3.19), expression (3.15) can be rewritten to the following form:

$$\begin{aligned} \mathbf{E_{FF}}(\theta,\varphi) &= E_\theta(\theta,\varphi)\mathbf{i_\theta} + E_\varphi(\theta,\varphi)\mathbf{i}_\varphi, \\ \mathbf{E_{FF}}(\theta,\varphi) &= B\mathbf{f}(\theta,\varphi) = B\left\{f_\theta(\theta,\varphi)\mathbf{i_\theta} + f_\varphi(\theta,\varphi)\mathbf{i}_\varphi\right\}, \end{aligned} \tag{3.20}$$

where B is the amplitude factor; $\mathbf{f}(\theta, \varphi)$ is the total radiation pattern of the radiating aperture; $f_\theta(\theta, \varphi)$ is the RP of the radiating aperture for θ component of $\mathbf{E_{FF}}(\theta, \varphi)$; and $f_\varphi(\theta, \varphi)$ is the RP of the radiating aperture for φ component of $\mathbf{E_{FF}}(\theta, \varphi)$:

$$f_\theta(\theta,\varphi) = \frac{(1+\cos\theta)}{2}\left(\cos\varphi N_x(\theta,\varphi) + \sin\varphi N_y(\theta,\varphi)\right); \tag{3.21}$$

$$f_\varphi(\theta,\varphi) = \frac{(1+\cos\theta)}{2}\left(-\sin\varphi N_x(\theta,\varphi) + \cos\varphi N_y(\theta,\varphi)\right). \tag{3.22}$$

For (R, α, β) and (x, y, z) coordinate systems, relationship between the unit vectors $\mathbf{i_R}$, $\mathbf{i_\alpha}$, $\mathbf{i_\beta}$ and $\mathbf{i_x}$, $\mathbf{i_y}$, $\mathbf{i_z}$ is defined by the following coordinate transformation matrix [1]:

$$\begin{pmatrix} \mathbf{i_R} \\ \mathbf{i_\alpha} \\ \mathbf{i_\beta} \end{pmatrix} = \begin{pmatrix} \cos\beta\sin\alpha & \sin\beta & \cos\beta\cos\alpha \\ \cos\alpha & 0 & -\sin\alpha \\ -\sin\alpha\sin\beta & \cos\beta & -\cos\alpha\sin\beta \end{pmatrix} \cdot \begin{pmatrix} \mathbf{i_x} \\ \mathbf{i_y} \\ \mathbf{i_z} \end{pmatrix} \tag{3.23}$$

In view of (3.10) and (3.23), one can write:

$$\rho\cos\gamma = (\mathbf{R}, \boldsymbol{\rho})/R = (\mathbf{i_R}, \boldsymbol{\rho}) = x\sin\alpha\cos\beta + y\sin\beta. \tag{3.24}$$

Since $\mathbf{i_R} = [\mathbf{i_\alpha}, \mathbf{i_\beta}]$, $(\mathbf{i_\alpha}, \mathbf{i_x}) = \cos\alpha$, $(\mathbf{i_\beta}, \mathbf{i_x}) = -\sin\alpha\sin\beta$, $(\mathbf{i_\alpha}, \mathbf{i_y}) = 0$, $(\mathbf{i_\beta}, \mathbf{i_y}) = \cos\beta$.

As well as the fact that for large size planar cophase apertures, the normal $\mathbf{n}$ can be considered constant throughout all aperture, vector transformations can be removed from the integral, and expression (3.12) takes the form:

$$\mathbf{E}_{\mathrm{FF}} = \frac{\mathrm{j}}{\lambda}\frac{\mathrm{e}^{-\mathrm{j}kR}}{R}\left\{\mathbf{e_x}(\alpha,\beta)N_x(\alpha,\beta) + \mathbf{e_y}(\alpha,\beta)N_y(\alpha,\beta)\right\}, \tag{3.25}$$

where the vector coefficients $\mathbf{e_x}(\alpha, \beta)$ and $\mathbf{e_y}(\alpha, \beta)$ have the following form:

$$\mathbf{e_x}(\alpha,\beta) = \frac{(\cos\alpha + \cos\beta)}{2}\mathbf{i}_\alpha - \frac{\sin\alpha\sin\beta}{2}\mathbf{i}_\beta; \tag{3.26}$$

$$\mathbf{e_y}(\alpha,\beta) = \frac{\sin\alpha\sin\beta}{2}\mathbf{i}_\alpha + \frac{(\cos\alpha + \cos\beta)}{2}\mathbf{i}_\beta; \tag{3.27}$$

$$N_x(\alpha,\beta) = \int_S a_x(x,y)\cdot \mathrm{e}^{ik(x\sin\alpha\cos\beta + y\sin\beta)}\mathrm{d}x\mathrm{d}y; \tag{3.28}$$

$$N_y(\alpha,\beta) = \int_S a_y(x,y)\cdot \mathrm{e}^{ik(x\sin\alpha\cos\beta + y\sin\beta)}\mathrm{d}x\mathrm{d}y. \tag{3.29}$$

Taking into account (3.26)–(3.29), expression (3.25) can be rewritten as follows:

$$\begin{aligned} \mathbf{E_{FF}}(\alpha,\beta) &= E_\alpha(\alpha,\beta)\mathbf{i}_\alpha + E_\beta(\alpha,\beta)\mathbf{i}_\beta, \\ \mathbf{E_{FF}}(\alpha,\beta) &= B\mathbf{f}(\alpha,\beta) = B\left\{f_\alpha(\alpha,\beta)\mathbf{i}_\alpha + f_\beta(\alpha,\beta)\mathbf{i}_\beta\right\}, \end{aligned} \tag{3.30}$$

where B is the amplitude factor; $\mathbf{f}(\alpha, \beta)$ is the radiating aperture total radiation pattern; $f_\alpha(\alpha, \beta)$ is the RP of radiating aperture for α component of $\mathbf{E_{FF}}(\alpha, \beta)$; and $f_\beta(\alpha, \beta)$ is the RP of radiating aperture for β component of $\mathbf{E_{FF}}(\alpha, \beta)$ vector

$$f_\alpha(\alpha,\beta) = \frac{(\cos\alpha + \cos\beta)}{2} N_x(\alpha,\beta) + \frac{\sin\alpha\sin\beta}{2} N_y(\alpha,\beta); \tag{3.31}$$

$$f_\beta(\alpha,\beta) = -\frac{\sin\alpha\sin\beta}{2} N_x(\alpha,\beta) + \frac{(\cos\alpha + \cos\beta)}{2} N_y(\alpha,\beta). \tag{3.32}$$

For plane apertures, the linear dimensions of which are much greater than the wavelength, when calculating the main and near side lobes of the radiation pattern, one can assume that the normal **n** to the aperture and the unit vector $\mathbf{i_R}$ of radius vector **R** are parallel to each other. Hence, in parentheses of expression (3.12), one can introduce the following approximation:

$$\mathbf{i}_R = -\mathbf{n} = \mathbf{i_z}. \tag{3.33}$$

In view of (3.33), expression (3.12) can be represented as:

$$\mathbf{E_{FF}} = \frac{\mathrm{j}}{\lambda}\frac{\mathrm{e}^{-\mathrm{j}kR}}{R}\int\limits_S \left[\mathbf{i_R}, \left[\mathbf{i}_z, \left[a_x\mathbf{i}_x + a_y\mathbf{i}_y\right]\right]\right]\mathrm{e}^{\mathrm{j}k\rho\cos\gamma}\mathrm{d}S \tag{3.34}$$

In this case, for the spherical coordinate system of the sighting point, the formulas for the radiation patterns $f_\theta(\theta,\varphi)$ and $f_\varphi(\theta,\varphi)$ are expressed as follows:

$$f_\theta(\theta,\varphi) = \cos\varphi N_x(\theta,\varphi) + \sin\varphi N_y(\theta,\varphi), \tag{3.35}$$

$$f_\varphi(\theta,\varphi) = \cos\theta\big(-\sin\varphi N_x(\theta,\varphi) + \cos\varphi N_y(\theta,\varphi)\big). \tag{3.36}$$

For the azimuth–elevation coordinate system of the sighting point, the expressions for the radiation patterns $f_\alpha(\alpha,\beta)$ and $f_\beta(\alpha,\beta)$ take the following form:

$$f_\alpha(\alpha,\beta) = \cos\beta N_x(\alpha,\beta) + \sin\alpha\sin\beta N_y(\alpha,\beta), \tag{3.37}$$

$$f_\beta(\alpha,\beta) = \cos\alpha N_y(\alpha,\beta). \tag{3.38}$$

Expressions for normalized amplitude radiation patterns can be written as:

$$\begin{aligned}
F_\theta(\theta,\varphi) &= |f_\theta(\theta,\varphi)|/|f_\theta(\theta,\varphi)|_{\max}\\
F_\varphi(\theta,\varphi) &= \left|f_\varphi(\theta,\varphi)\right|/\left|f_\varphi(\theta,\varphi)\right|_{\max}\\
F_\alpha(\alpha,\beta) &= |f_\alpha(\alpha,\beta)|/|f_\alpha(\alpha,\beta)|_{\max}\\
F_\beta(\alpha,\beta) &= \left|f_\beta(\alpha,\beta)\right|/\left|f_\beta(\alpha,\beta)\right|_{\max}
\end{aligned}$$

3.3 Analysis of Calculating Expressions Based on the Solution of a Homogeneous Wave Equation

At present, methods for calculating the far field of the antenna from the measured complex amplitude (i.e., amplitude and phase) of the vector **E** on the scan surface are based on the statement that this field can be represented as a superposition of elementary waves (modes) with planar, cylindrical, or spherical phase front. Such a representation is called an eigenmode expansion. The amplitudes and phases of eigenmodes are calculated by measuring the tangential component of the vector **E** on the scan surface, namely in the case of plane wave expansion—on the plane, in the case of the cylindrical wave expansion—on the side surface of the circular cylinder covering the AUT, in the case of spherical wave expansion—on the sphere encompassing the AUT. Moreover, the distance of these surfaces from the studied antenna is not critical. The data measured in the near field of the antenna is converted into the sum (or spectrum) of eigenwaves, in coordinate system, one of the coordinate surfaces of which has the same shape as scan surface. The probe's radiation pattern, which is considered a priori known, is also expressed through the expansion in eigenwaves determined by the same wave numbers, which allows accomplishing probe correction. Having this data, it is possible to calculate the radio characteristics of the AUT.

Let us consider in more detail the use of the eigenwave expansion method using planar expansion as an example. In an isotropic, homogeneous, lossless medium, free of foreign currents and charges, Maxwell equations for harmonic electromagnetic field vectors can be replaced with Helmholtz equation for complex amplitude of these vectors [1]. For the complex amplitude of the electric field vector **E**, these equations have the following form:

$$\nabla^2\mathbf{E} + k^2\mathbf{E} = 0, \tag{3.39}$$

$$\nabla\mathbf{E} = 0, \tag{3.40}$$

$$k^2 = \omega^2\mu\varepsilon, \tag{3.41}$$

where k is the wavenumber and ε, μ are the absolute dielectric permittivity and magnetic permeability of the medium accordingly.

Expression (3.39) for the complex amplitude of the vector **E** is a partial differential equation and has a set of solutions. One of the particular solutions to this equation is a plane homogeneous wave propagating in the direction specified by the wave vector k and can be written in the following form:

$$\mathbf{E}(\mathbf{r}) = \mathbf{A}(\mathbf{k})\mathrm{e}^{-\mathrm{j}(\mathbf{kr})}, \tag{3.42}$$

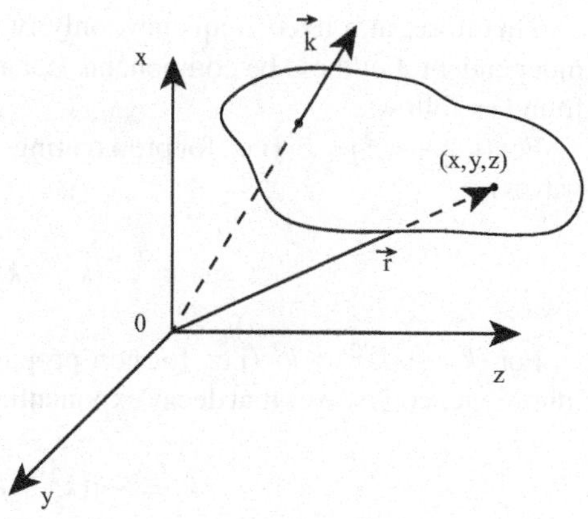

Fig. 3.6 Instantaneous position of the plane wave phase front and the orientation of the vectors **k** and **r** relative to this front

where **r** is the radius vector of the sighting point at which vector **E** is calculated; **A**(**k**) is the complex vector amplitude of the plane wave that represents amplitude and phase dependence of this wave on the propagation direction.

Figure 3.6 shows the instantaneous position of the plane wave phase front and the orientation of the vectors **k** and **r** relative to this front (**k** is perpendicular to the plane of the phase front; **r** is directed to an arbitrary sighting point on the surface of the phase front).

In the general case, all vector values in expression (3.42) have three independent components. In a Cartesian coordinate system, these vectors can be written as follows:

$$\mathbf{r} = r_x\mathbf{i_x} + r_y\mathbf{i}_y + r_z\mathbf{i}_z = x\mathbf{i_x} + y\mathbf{i}_y + z\mathbf{i}_z, \tag{3.43}$$

where r_x, r_y, r_z are the projections of the vector **r** on the coordinate axes; *x*, *y*, *z* are the coordinates of the sighting point at which the vector **E** is calculated; and $\mathbf{i_x}$, $\mathbf{i_y}$, $\mathbf{i_z}$ are the unit vectors of a Cartesian coordinate system.

$$\mathbf{k} = k_x\mathbf{i_x} + k_y\mathbf{i}_y + k_z\mathbf{i}_z \tag{3.44}$$

where k_x, k_y, k_z are projections of the vector **k** on the coordinate axes.

$$\mathbf{A}(\mathbf{k}) = A_x\left(k_x, k_y, k_z\right)\mathbf{i_x} + A_y\left(k_x, k_y, k_z\right)\mathbf{i_y} + A_z\left(k_x, k_y, k_z\right)\mathbf{i_z} \tag{3.45}$$

$$\mathbf{E}(\mathbf{r}) = \mathbf{E}(x, y, z) = E_x(x, y, z)\mathbf{i_x} + E_y(x, y, z)\mathbf{i_y} + E_z(x, y, z)\mathbf{i_z} \tag{3.46}$$

However, for the conditions under consideration, when expressions (3.39)–(3.41) are valid, not all components of the vectors **k**, **A**(**k**), and **E** are independent. Let us find the independent components of the vector **k**. Substituting (3.42) into (3.39), one can get $\mathbf{kk} = k^2$. Hence,

$$k_x^2 + k_y^2 + k_z^2 = \omega^2\mu\varepsilon \tag{3.47}$$

Therefore, at a fixed frequency, only two components of the wave vector **k** are independent. Let these be components k_x and k_y, then the third component k_z can be found as follows:

For $({k_x}^2 + {k_y}^2) \leq k^2$ (i.e., for propagating electromagnetic fields—electromagnetic waves),

$$k_z = \left(k^2 - k_x^2 - k_y^2\right)^{0.5} \tag{3.48}$$

For $({k_x}^2 + {k_y}^2) > k^2$ (i.e., for non-propagating reactive electromagnetic fields—inhomogeneous waves that decay exponentially, at least along one of the coordinates),

$$k_z = -\mathrm{j}\left(k_x^2 + k_y^2 - k^2\right)^{0.5}. \tag{3.49}$$

Thus, for a plane electromagnetic wave in the medium under consideration,

$$\mathbf{k} = \mathbf{k}\left(k_x, k_y\right) = k_x \mathbf{i_x} + k_y \mathbf{i}_y + \left(k^2 - k_x^2 - k_y^2\right)^{0.5} \mathbf{i}_z \tag{3.50}$$

Let us find independent components of the vector **A(k)**. Substituting (3.42) into (3.40), one can obtain **k·A(k)** = 0.

$$k_x A_x\left(k_x, k_y\right) + k_y A_y\left(k_x, k_y\right) + k_z A_z\left(k_x, k_y\right) = 0. \tag{3.51}$$

Thus, for any direction specified by the wave vector **k**, only two components of the vector amplitude **A(k)** are independent. Let these be A_x and A_y. Then,

$$A_z\left(k_x, k_y\right) = -\left[k_x A_x\left(k_x, k_y\right) + k_y A_y\left(k_x, k_y\right)\right] / k_z \tag{3.52}$$

Hence, for a plane wave in the medium under consideration,

$$\mathbf{A(k)} = \mathbf{A}\left(k_x, k_y\right) = A_x\left(k_x, k_y\right)\mathbf{i_x} + A_y\left(k_x, k_y\right)\mathbf{i_y} - \frac{k_x A_x\left(k_x, k_y\right) + k_y A_y\left(k_x, k_y\right)}{\left(k^2 - k_x^2 - k_y^2\right)^{0.5}} \mathbf{i_z} \tag{3.53}$$

Let us find independent components of the vector **E**. Rewriting Eq. (3.40) in the Cartesian coordinate system, one can obtain:

$$\frac{\partial E_x}{\partial x} + \frac{\partial E_y}{\partial y} + \frac{\partial E_z}{\partial z} = 0. \tag{3.54}$$

Thus, if two components of the vector **E** are known, then the third one can be found from Eq. (3.54). Let the independent components be E_x and E_y. All further calculations will be performed for the independent components of the vectors **k**, **A(k)**, and **E**. Since the field in the form of a plane wave with arbitrary parameters is a particular solution of Eq. (3.39), the general solution to this equation is the sum (superposition) of fields of the form (3.42) with arbitrary amplitudes, phases, and propagation directions. Such a solution can be written as an integral over the independent parameters of plane waves, which are the amplitude, phase, and direction of propagation [6]:

$$\mathbf{E}(x, y, z) = \int_{-\infty}^{\infty}\int_{-\infty}^{\infty} \mathbf{A}(k_x k_y)\exp(-\mathrm{j}(k_x x + k_y y + k_z z))\mathrm{d}k_x \mathrm{d}k_y. \quad (3.55)$$

In a Cartesian coordinate system, vector $\mathbf{E}(x, y, z)$ components are expressed as follows:

$$E_x(x, y, z) = \int_{-\infty}^{\infty}\int_{-\infty}^{\infty} A_x(k_x k_y)\exp(-\mathrm{j}(k_x x + k_y y + k_z z))\mathrm{d}k_x \mathrm{d}k_y. \quad (3.56)$$

$$E_y(x, y, z) = \int_{-\infty}^{\infty}\int_{-\infty}^{\infty} A_y(k_x k_y)\exp(-\mathrm{j}(k_x x + k_y y + k_z z))\mathrm{d}k_x \mathrm{d}k_y. \quad (3.57)$$

$$E_z(x, y, z) = \int_{-\infty}^{\infty}\int_{-\infty}^{\infty} A_z(k_x k_y)\exp(-\mathrm{j}(k_x x + k_y y + k_z z))\mathrm{d}k_x \mathrm{d}k_y. \quad (3.58)$$

Expression (3.55) describes the field at any point in space. In this expression, the parameters x, y, z are the Cartesian coordinates of the sighting point, and the function $A\ (k_x\ k_y)$ describes the amplitudes and phases of the plane waves passing through this point, which propagate in the directions determined by the independent components of the wave vector. This function is called the *angular spectrum of plane waves or plane wave spectrum (PWS)* of the **E** field. The name "angular spectrum" reflects the relationship of the components k_x, k_y of the wave vector k with the propagation angles of plane waves forming the spectrum, as well as the formal similarity of these components to the frequency spectrum of the Fourier transform in the time domain. By the same analogy, the variables k_x, k_y are sometimes called the spatial frequencies of the variables x and y. Since the amplitude and initial phase of the plane wave vector E during propagation in the lossless medium remain unchanged, then the PWS of the field excited by the antenna can be found in any region of space.

Let us show that for the case under consideration, the amplitude function $\mathbf{A}(k_x\ k_y)$ can be found analytically if the values of the tangential component $\mathbf{E}_\tau$ of the vector $\mathbf{E}$ on a plane parallel to one of the coordinate surfaces of a Cartesian coordinate system are known. Let the radiating aperture be located in $z = 0$ plane of the Cartesian coordinate system $OXYZ$, and the plane in which the values of the tangential component E_τ are determined is the $z = d$ plane. Figure 3.7 shows radiating aperture plane location. Let us introduce notations $E_{\tau x}(x, y, d)$ and $E_{\tau y}(x, y, d)$ for the orthogonal projections of the vector $\mathbf{E}_\tau$ onto the plane $z = d$. Substituting $z = d$ into expressions (3.56) and (3.57), one can obtain:

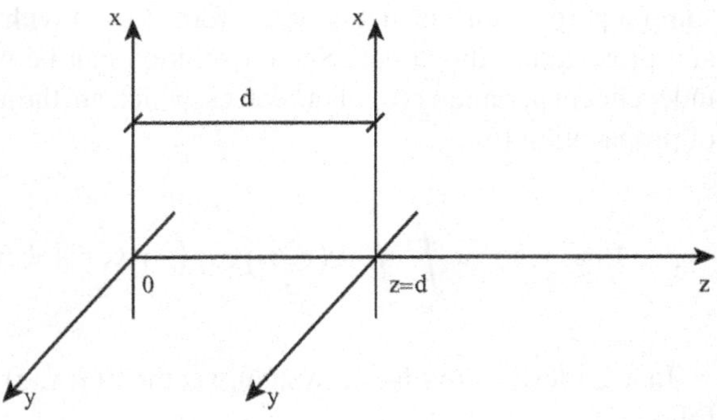

Fig. 3.7 Radiating aperture location

$$E_{\tau x}(x, y, d) = \int_{-\infty}^{\infty}\int_{-\infty}^{\infty} A_x(k_x k_y)\exp(-jk_z d)\exp(-j(k_x x + k_y y))dk_x dk_y. \quad (3.59)$$

$$E_{\tau y}(x, y, d) = \int_{-\infty}^{\infty}\int_{-\infty}^{\infty} A_y(k_x k_y)\exp(-jk_z d)\exp(-j(k_x x + k_y y))dk_x dk_y. \quad (3.60)$$

Expressions (3.59), (3.60) analysis shows that A_x and A_y are two-dimensional Fourier transforms of the $E_{\tau x}$(x,y,d) and $E_{\tau y}$(x,y,d) respectively. Therefore, the inverse two-dimensional Fourier transform formula is applicable:

$$A_x(k_x k_y) = (\exp(jk_z d)/4\pi^2)\int_{-\infty}^{\infty}\int_{-\infty}^{\infty} E_{\tau x}(x, y, d)\exp(j(k_x x + k_y y))dxdy. \quad (3.61)$$

$$A_y(k_x k_y) = (\exp(jk_z d)/4\pi^2)\int_{-\infty}^{\infty}\int_{-\infty}^{\infty} E_{\tau y}(x, y, d)\exp(j(k_x x + k_y y))dxdy. \quad (3.62)$$

Substituting A_x and A_y from (3.61) and (3.62) into (3.56) and (3.57), one can obtain expressions that allow using the known values of the orthogonal vector $\mathbf{E}_\tau$ projections on the plane $z = d$, to calculate the components of vector **E** at any point of the half-space $z \geq 0$. If it is necessary to diagnose amplitude–phase distribution in radiating aperture ($z = 0$) based on measured values of $E_{\tau x}(x, y, d)$ and $E_{\tau y}(x, y, d)$, then the calculation formulas get the following form:

$$E_x(x, y, 0) = \int_{-\infty}^{\infty}\int_{-\infty}^{\infty} A_x(k_x k_y)\exp(-j(k_x x + k_y y))dk_x dk_y, \quad (3.63)$$

$$E_y(x, y, 0) = \int_{-\infty}^{\infty}\int_{-\infty}^{\infty} A_y(k_x k_y)\exp(-\mathrm{j}(k_x x + k_y y))\mathrm{d}k_x \mathrm{d}k_y, \tag{3.64}$$

where $E_x(x, y, 0)$ and $E_y(x, y, 0)$ are the electric field strength vector components at the aperture; k_z is defined by expression (3.48).

In general, calculating the quadruple integrals obtained by substituting (3.61) and (3.62) into (3.56) and (3.57) is not an easy task. However, if it is necessary to find only the external radio characteristics of the antenna, a significant simplification can be made.

Since these characteristics are defined for the far-field region of the antenna, the stationary phase method (saddle point method) can be used to calculate the above integrals. The application of this method shows that the major contribution to the value of **E** at an arbitrary sighting point in the far-field region comes from that plane wave from the entire spectrum in which propagation direction coincides with the direction of the radius vector **r** of this point. If we turn to Fig. 3.4, then for this case vectors k and r are parallel to each other. As a result, the expression for the far-field **E** is significantly simplified and takes the following form [5, 6]

$$\mathbf{E}_{\mathbf{FF}}(\mathbf{r}) = \mathrm{j}\frac{2\pi}{r}k_{z0}\mathbf{A}(\mathbf{k}_0)\exp(-\mathrm{j}(\mathbf{k}_0\mathbf{r})) \tag{3.65}$$

where $\mathbf{k_0} = k\mathbf{i_r}$; $\mathbf{i_r}$ is the unit vector in the direction of $\mathbf{r}$; and $\mathbf{r} = r\,\mathbf{i_r}$.

The scalar product $(\mathbf{k_0 r})$ in this case is equal to $k{\cdot}r$. In a Cartesian coordinate system, vector $\mathbf{E_{FF}}(\mathbf{r})$ components are expressed as follows:

$$E_{x\mathrm{FF}}(x, y, z) = \mathrm{j}2\pi k_{z0}\exp(-\mathrm{j}(kr)/r)A_x(k_{x0}, k_{y0}), \tag{3.66}$$

$$E_{y\mathrm{FF}}(x, y, z) = \mathrm{j}2\pi k_{z0}\exp(-\mathrm{j}(kr)/r)A_y(k_{x0}, k_{y0}). \tag{3.67}$$

The component $E_{z\mathrm{FF}}(x, y, z)$ could be found from Eq. (3.54). However, in this case it is easier to find $A_z(k_{x0}\ k_{y0})$ by an expression similar to (3.52):

$$A_z(k_{x0}, k_{y0}) = -[k_{x0}A_x(k_{x0}, k_{y0}) + k_{y0}A_y(k_{x0}, k_{y0})]/k_{z0} \tag{3.68}$$

and then calculate $E_{z\mathrm{FF}}(x, y, z)$ using expressions similar to (3.66), (3.67)

$$E_{z\mathrm{FF}}(x, y, z) = \mathrm{j}2\pi k_{z0}\exp(-\mathrm{j}(kr)/r)A_z(k_{x0}, k_{y0}). \tag{3.69}$$

In the far-field region, vector **E** lies entirely in a plane perpendicular to the wave propagation direction. Therefore to calculate the antenna radiation pattern, it is more convenient to express it in the form of components $E_\theta(r, \theta, \varphi)$ and $E_\varphi(r, \theta, \varphi)$ of a spherical coordinate system or in the form of components $E_\alpha(r, \alpha, \beta)$ and $E_\beta(r, \alpha, \beta)$ of the azimuth–elevation coordinate system depicted in Figs. 3.2 and 3.3, respectively. In contrast to Cartesian one, in these coordinate systems vector **E** has not three, but only two components, since its axial component E_r is equal to 0. The relationship of the components E_θ, E_φ and E_α, E_β of vector **E** with the components $E_{x\text{FF}}$, $E_{y\text{FF}}$, $E_{z\text{FF}}$ of this vector is defined by the coordinate transformation matrices (3.13) and (3.23)

$$E_\theta = \cos\theta \cos\varphi E_{x\text{FF}} + \cos\theta \sin\varphi E_{y\text{FF}} - \sin\theta E_{z\text{FF}} \tag{3.70}$$

$$E_\varphi = -\sin\varphi E_{x\text{FF}} + \cos\varphi E_{y\text{FF}} \tag{3.71}$$

$$E_\alpha = \cos\alpha E_{x\text{FF}} - \sin\alpha E_{z\text{FF}} \tag{3.72}$$

$$E_\beta = -\sin\alpha \sin\beta E_{x\text{FF}} + \cos\beta E_{y\text{FF}} - \cos\alpha \sin\beta E_{z\text{FF}} \tag{3.73}$$

The same matrices allow to derive expressions for $\mathbf{i_r}$,

$$\mathbf{i_r}(\theta, \varphi) = \sin\theta \cos\varphi \mathbf{i_x} + \sin\theta \sin\varphi \mathbf{i}_y + \cos\theta \mathbf{i}_z, \tag{3.74}$$

$$\mathbf{i_r}(\alpha, \beta) = \cos\beta \sin\alpha \mathbf{i_x} + \sin\beta \mathbf{i}_y + \cos\beta \cos\alpha \mathbf{i}_z. \tag{3.75}$$

Taking into account (3.65), the expressions for the components of the vector $\mathbf{k_0}$ are as follows:

$$\begin{aligned} k_{x0}(\theta, \varphi) &= k \sin\theta \cos\varphi \\ k_{y0}(\theta, \varphi) &= k \sin\theta \sin\varphi \\ k_{z0}(\theta, \varphi) &= k \cos\theta \end{aligned} \tag{3.76}$$

$$\begin{aligned} k_{x0}(\alpha, \beta) &= k \cos\beta \sin\alpha \\ k_{y0}(\alpha, \beta) &= k \sin\beta \\ k_{z0}(\alpha, \beta) &= k \cos\beta \cos\alpha \end{aligned} \tag{3.77}$$

Substituting (3.76) into expressions (3.66), (3.67), (3.69) for $E_{x\text{FF}}$, $E_{y\text{FF}}$, $E_{z\text{FF}}$, and the results into (3.70), (3.71), one can obtain:

$$E_\theta = \text{j}2\pi k \exp(-\text{j}(kr)/r)\left(A_x\left(k_{x0}, k_{y0}\right) \cos\varphi + A_y\left(k_{x0}, k_{y0}\right) \sin\varphi\right) \tag{3.78}$$

$$E_\varphi = \text{j}2\pi k \exp(-\text{j}(kr)/r)\left(-A_x\left(k_{x0}, k_{y0}\right) \sin\varphi + A_y\left(k_{x0}, k_{y0}\right) \cos\varphi\right) \cos\theta \tag{3.79}$$

Substituting (3.77) into expressions (3.66), (3.67), (3.69) for $E_{x\text{FF}}$, $E_{y\text{FF}}$, $E_{z\text{FF}}$, and the results into (3.72), (3.73), one can obtain:

$$E_\alpha = \text{j}2\pi k \exp(-\text{j}(kr)/r)\left(A_x\left(k_{x0}, k_{y0}\right) \cos\beta + A_y\left(k_{x0}, k_{y0}\right) \sin\alpha \sin\beta\right) \tag{3.80}$$

$$E_\beta = -\mathrm{j}2\pi k \exp(-\mathrm{j}(kr)/r) A_y\left(k_{x0}, k_{y0}\right) \cos\alpha \tag{3.81}$$

Expressions (3.78), (3.79) are the basis for calculating antenna RP in the spherical coordinate system, and expressions (3.80), (3.81) are the basis for calculating antenna RP in the azimuth–elevation coordinate system.

For the spherical coordinate system,

$$\mathbf{E}_{\mathbf{FF}}(\theta, \varphi) = E_\theta(\theta, \varphi)\mathbf{i}_\theta + E_\varphi(\theta, \varphi)\mathbf{i}_\varphi = B\mathbf{f}(\theta, \varphi), \tag{3.82}$$

where B is a constant factor and $\mathbf{f}(\theta, \varphi)$ is the total complex vector antenna radiation pattern, which is calculated as follows:

$$\mathbf{f}(\theta, \varphi) = f_\theta(\theta, \varphi)\mathbf{i}_\theta + f_\varphi(\theta, \varphi)\mathbf{i}_\varphi, \tag{3.83}$$

where $f_\theta(\theta, \varphi)$ is the RP of the complex amplitude of θ component of the vector $\mathbf{E}_{\mathbf{FF}}(\theta, \varphi)$ and $f_\varphi(\theta, \varphi)$ is the RP of the complex amplitude of φ component of the vector $\mathbf{E}_{\mathbf{FF}}(\theta, \varphi)$.

For a linearly polarized wave, the components $f_\theta(\theta, \varphi)$ and $f_\varphi(\theta, \varphi)$ are in-phase and orthogonal; therefore, the total amplitude RP ($|f(\theta, \varphi)|$) is equal to:

$$|f(\theta, \varphi)| = \left(|f_\theta(\theta, \varphi)|^2 + \left|f_\varphi(\theta, \varphi)\right|^2\right)^{0.5} \tag{3.84}$$

The components $f_\theta(\theta, \varphi)$ and $f_\varphi(\theta, \varphi)$ are calculated as follows:

$$f_\theta(\theta, \varphi) = A_x\left(k_{x0}, k_{y0}\right)\cos\varphi + A_y\left(k_{x0}, k_{y0}\right)\sin\varphi \tag{3.85}$$

$$f_\varphi(\theta, \varphi) = \left(-A_x\left(k_{x0}, k_{y0}\right)\sin\varphi + A_y\left(k_{x0}, k_{y0}\right)\cos\varphi\right)\cos\theta \tag{3.86}$$

In order to make the dependence of the parameters $A_x(k_{x0}\ k_{y0})$ and $A_y(k_{x0}\ k_{y0})$ on the angles θ, φ more explicit in (3.85) and (3.86), they can be written as follows:

$$A_x\left(k_{x0}k_{y0}\right) = A_x(\theta, \varphi) = \left(\exp(\mathrm{j}k_{z0}d)/4\pi^2\right) \int\limits_{-\infty}^{\infty}\int\limits_{-\infty}^{\infty} E_{\tau x}(x, y, d)\exp(\mathrm{j}k(x\sin\theta\cos\varphi + y\sin\theta\sin\varphi))\mathrm{d}x\mathrm{d}y, \tag{3.87}$$

$$A_y\left(k_{x0}k_{y0}\right) = A_y(\theta, \varphi) = \left(\exp(\mathrm{j}k_{z0}d)/4\pi^2\right)$$

$$\int_{-\infty}^{\infty}\int_{-\infty}^{\infty} E_{\tau y}(x, y, d)\exp(jk(x\sin\theta\cos\varphi + y\sin\theta\sin\varphi))dxdy. \tag{3.88}$$

Normalized amplitude radiation patterns in relative units are expressed as follows:

$$F_\theta(\theta, \varphi) = |f_\theta(\theta, \varphi)|/|f_\theta(\theta, \varphi)|_{\max} \tag{3.89}$$

$$F_\varphi(\theta, \varphi) = \left|f_\varphi(\theta, \varphi)\right|/\left|f_\varphi(\theta, \varphi)\right|_{\max} \tag{3.90}$$

The total power RP ($|f(\theta, \varphi)|^2$) is equal to:

$$|f(\theta, \varphi)|^2 = |f_\theta(\theta, \varphi)|^2 + \left|f_\varphi(\theta, \varphi)\right|^2 \tag{3.91}$$

For the azimuth–elevation coordinate system,

$$\mathbf{E}_{\mathbf{FF}}(\alpha, \beta) = E_\alpha(\alpha, \beta)\mathbf{i}_\alpha + E_\beta(\alpha, \beta)\mathbf{i}_\beta = B\mathbf{f}(\alpha, \beta), \tag{3.92}$$

where B is the constant factor and $\mathbf{f}(\alpha, \beta)$ is the total complex vector antenna RP, which is calculated as follows:

$$\mathbf{f}(\alpha, \beta) = \left\{f_\alpha(\alpha, \beta)\mathbf{i}_\alpha + f_\beta(\alpha, \beta)\mathbf{i}_\beta\right\} \tag{3.93}$$

where $f_\alpha(\alpha, \beta)$ is the RP of the complex amplitude of the α component of $\mathbf{E}_{\mathbf{FF}}(\alpha, \beta)$ and $f_\beta(\alpha, \beta)$ is the RP of the complex amplitude of the β component of $\mathbf{E}_{\mathbf{FF}}(\alpha, \beta)$.

For a linearly polarized wave, the components $f_\alpha(\alpha, \beta)$ and $f_\beta(\alpha, \beta)$ are in-phase and orthogonal; therefore, total amplitude RP ($|f(\alpha, \beta)|$) is equal to:

$$|f(\alpha, \beta)| = \left(|f_\alpha(\alpha, \beta)|^2 + \left|f_\beta(\alpha, \beta)\right|^2\right)^{0.5}. \tag{3.94}$$

The components $f_\alpha(\alpha, \beta)$ and $f_\beta(\alpha, \beta)$ are calculated as follows:

$$f_\alpha(\alpha, \beta) = A_x\left(k_{x0}, k_{y0}\right)\cos\beta + A_y\left(k_{x0}, k_{y0}\right)\sin\alpha\sin\beta. \tag{3.95}$$

$$f_\beta(\alpha, \beta) = A_y\left(k_{x0}, k_{y0}\right)\cos\alpha. \tag{3.96}$$

In order to make the dependence of the parameters $A_x(k_{x0}\ k_{y0})$ and $A_y(k_{x0}\ k_{y0})$ on the angles α, β more explicit in (3.95), (3.96), they can be written in the following form:

$$A_x\left(k_{x0}k_{y0}\right) = A_x(\alpha, \beta) = \left(\exp(jk_z d)/4\pi^2\right)$$

$$\int_{-\infty}^{\infty}\int_{-\infty}^{\infty} E_{\tau x}(x, y, d)\exp(\mathrm{j}k(x \sin\alpha \cos\beta + y \sin\beta))\mathrm{d}x\mathrm{d}y, \quad (3.97)$$

$$A_y\left(k_{x0}k_{y0}\right) = A_y(\alpha, \beta) = \left(\exp(\mathrm{j}k_z d)/4\pi^2\right)$$
$$\int_{-\infty}^{\infty}\int_{-\infty}^{\infty} E_{\tau y}(x, y, d)\exp(\mathrm{j}k(x \sin\alpha \cos\beta + y \sin\beta))\mathrm{d}x\mathrm{d}y. \quad (3.98)$$

Total power RP ($|f(\alpha, \beta)|^2$) is equal to:

$$|f(\alpha, \beta)|^2 = |f_\alpha(\alpha, \beta)|^2 + \left|f_\beta(\alpha, \beta)\right|^2 \quad (3.99)$$

Eventually, it can be noted that expressions (3.85), (3.86) coincide with expressions (3.35) and (3.36), and expressions (3.95), (3.96) coincide with expressions (3.37) and (3.38).

3.4 Discrete Analogs of Integral Transforms

Let us take a closer look on integral transforms that connect the complex amplitude values of the tangential component $\mathbf{E}_\tau$ orthogonal projections on the scan plane with the plane wave spectrum (PWS) of vector $\mathbf{E}$. If one omits constant factors, then the two-dimensional integral Fourier transform determining this relationship can be written in the following form:

$$A_{x,y}\left(k_x k_y\right) = \int_{-\infty}^{\infty}\int_{-\infty}^{\infty} E_{\tau x,y}(x, y)\exp\left(\mathrm{j}\left(k_x x + k_y y\right)\right)\mathrm{d}x\mathrm{d}y, \quad (3.100)$$

where $E_{\tau x,y}(x, y)$ is the function that specifies the orthogonal component values of the complex amplitude $\mathbf{E}_\tau$ on the scan plane; x, y are the Cartesian coordinates of the point at the scan plane; k_x, k_y are the orthogonal components of the wave vector $\mathbf{k}$ of plane waves included in PWS (generalized angular coordinates of the sighting point in the far-field region of the AUT); and $A_{x,y}$ (k_x, k_y) is angular plane wave spectrum.

When taking measurements, the scan area cannot be infinitely large. Let us denote real scan area limits in which the values of the function $E_{\tau x,y}(x, y)$ are defined, by values $\pm X$, $\pm Y$

$$-X \le x \le X, -Y \le y \le Y \quad (3.101)$$

Then, the transverse dimensions of the scan area are $2X \times 2Y$, and the expression for PWS calculation takes the form:

$$A_{x,y}(k_x k_y) = \int_{-X}^{X}\int_{-Y}^{Y} E_{\tau x,y}(x, y)\exp[\mathrm{j}(k_x x + k_y y)]\mathrm{d}x\mathrm{d}y. \tag{3.102}$$

Since measurements can be carried out only at a finite number of points on the scanning surface, the integral Fourier transform must be replaced by its discrete analog, which is called the "discrete Fourier transform" (DFT). In the case under consideration, the DFT can be used because the near-field and the angular spectrum of plane waves are functions with bounded spectrum. Rather, they are functions with an "almost bounded spectrum" or "almost bandlimited" functions. The DFT corresponding to the integral transform (3.102) can be written in the following form (assuming that the variable x varies from 0 to $2X$ and the variable y from 0 to $2Y$):

$$A_{x,y}(q\Delta k_x, l\Delta k_y) = \sum_{m}^{M-1}\sum_{n}^{N-1} E_{\tau x,y}(m\Delta x, n\Delta y) \cdot \exp[\mathrm{j}(q\Delta k_x m\Delta x + l\Delta k_y n\Delta y)] \tag{3.103}$$

where $E_{\tau x,y}(m\Delta x, n\Delta y)$ are the measured values of the mutually perpendicular components of the complex amplitude of the vector $\mathbf{E}_\tau$ on the scan surface at points with coordinates $x = m\Delta x$, $y = n\Delta y$.

- Δx is the x-axis step, i.e., the distance along the x-axis between the points at which the function $E_{\tau x,y}(m\Delta x, n\Delta y)$ samples are measured;
- m is an integer variable that determines the position of the measurement point on the x-axis;
- M is the number of the function $E_{\tau x,y}(m\Delta x, n\Delta y)$ samples along the x-axis;
- Δy is the y-axis step, i.e., the distance along the y-axis between the points at which the samples of the function $E_{\tau x,y}(m\Delta x, n\Delta y)$ are measured;
- n is an integer variable that determines the position of the measurement point on the y-axis;
- N is the number of samples of the function $E_{\tau x,y}(m\Delta x, n\Delta y)$ along the y-axis;
- Δk_x is the step of k_x values, which determines the angular position of the sighting point in the far-field region of the AUT;
- q is an integer variable that determines the sighting point position in the far-field region of the AUT with respect to the variable k_x;
- Δk_y is the step of k_y values, which determines the sighting point angular position in the far-field region of the AUT;
- l is an integer variable that determines the sighting point position in the far-field region of the AUT with respect to the variable k_y.

$A_{x,y}(q\Delta k_x, l\Delta k_y)$ are the mutually perpendicular component calculated values of the PWS of the vector E in far-field region of the AUT at the sighting point with generalized angular coordinates $k_x = q{\cdot}\Delta k_x$, $k_y = l{\cdot}\Delta k_y$.

Values X, Y, M, N, Δx, and Δy are related as follows:

$$M = 2X/\Delta x + 1, N = 2Y/\Delta y + 1 \quad (3.104)$$

The DFT is a periodic function; therefore, $A_{x,y}(q\Delta k_x, l\Delta k_y)$ can serve as an analog of $A_{x,y}(k_x, k_y)$ in a limited range of k_x and k_y values. The DFT period is specified by the exponent superscript in expression (3.103). Hence, the upper and lower limits of the possible values of k_x turn out to be $\pm\pi/\Delta x$, and the upper and lower limits of the possible k_y values turn out to be $\pm\pi/\Delta y$. Let us introduce special notation for the limiting values of the spatial variables k_x and k_y and for the number of samples of the function $A_{kx,ky}(q\Delta k_x, l\Delta k_y)$. Then,

$K_{xm} = \pm\pi/\Delta$ are the limits of the possible k_x values; $K_{ym} = \pm\pi/\Delta y$ are the limits of the possible k_y values; Q is the number of samples of the function $A_{kx,ky}(q\Delta k_x, l\Delta k_y)$ with respect to the variable k_x; and L is the number of samples of the function $A_{kx,ky}(q\Delta k_x, l\Delta k_y)$ with respect to the variable k_y.

Then,

$$-K_{xm} \le k_x \le K_{xm}, -K_{ym} \le k_y \le K_{ym} \quad (3.105)$$

The following relationship exists between Q, L, K_{xm}, K_{ym}, Δk_x, and Δk_y:

$$Q = (2K_{xm}/\Delta k_x) + 1 = ((2\pi/\Delta x)/\Delta k_x) + 1 \quad (3.106)$$

$$L = \left(2K_{ym}/\Delta k_y\right) + 1 = \left((2\pi/\Delta y)/\Delta k_y\right) + 1 \quad (3.107)$$

Hence, $\Delta k_x\ \Delta x = 2\pi/(Q-1)$ and $\Delta k_y\ \Delta y = 2\pi/(L-1)$. Substituting the values of the products $(\Delta k_x\ \Delta x)$ and $(\Delta k_y\ \Delta y)$ into exponent superscript of the expression (3.103), one can obtain:

$$\mathbf{A_{x,y}}\left(\mathbf{q\Delta k_x, l\Delta k_y}\right) = \sum_{\mathbf{m}}^{\mathbf{M-1}}\sum_{\mathbf{n}}^{\mathbf{N-1}} \mathbf{E_{\tau x,y}(m\Delta x, n\Delta y)} \cdot \mathbf{exp[j2\pi((q \cdot m/(Q-1) + l \cdot n/(L-1))]}. \quad (3.108)$$

Since the quantities Δk_x, Δk_y Δx, and Δy are chosen before the calculations, the expression (3.108) can be written in the following compact form

$$\mathbf{A_{x,y}(q, l)} = \sum_{\mathbf{m}}^{\mathbf{M-1}}\sum_{\mathbf{n}}^{\mathbf{N-1}} \mathbf{E_{\tau x,y}(m, n)} \cdot \mathbf{exp[j2\pi((q \cdot m/(Q-1) + l \cdot n/(L-1))]}. \quad (3.109)$$

Planar measurement can be divided into the following steps:

- Determining the scan area size along the x- and y-axes, the Δx and Δy step values and the number of the function $E_{\tau x,y}(m\Delta x, n\Delta y)$ samples along x- and y-axes;
- The results measuring and storing;

- An array of experimental data preliminary processing;
- PWS calculating;
- 3D RP of the AUT calculating;
- Plotting 2D RPs in given section, calculating directivity, gain, phase, RP, etc.

When determining the boundaries of the scan area (X and Y values), as a rule, they are guided by the following, field-proven criterion—the amplitude of the field strength at the boundary of the scan area should be 35–40 dB below its maximum value on this plane. The values K_{xm} and K_{ym}, as a rule, are chosen equal to the maximum possible value of the spatial frequency of the PWS for the region of existence of propagating EM waves (i.e., $2\pi/\lambda$). The choice of $K_{xm} = K_{ym} = 2\pi/\lambda$ determines the minimum step values of the spatial variables x and y when DFT is used: $\Delta x = \Delta y = \lambda/2$.

In modern measurement facilities, continuous probe movement is used. The measurement of $|E_{\tau x}|$, $\varphi_{\tau x}$ takes place at discrete points, when the coordinate controller sends a trigger signal to the receiver. Since the sweep time of the receiver is not infinitesimal, and the speed of the probe is not absolutely stable, the measurements at each frequency are slightly shifted in space. A way to partially compensate for this phenomenon is to use a reverse sweep on even rows (columns). Nevertheless, the spatial distributions of $|E_{\tau x}|$ and $\varphi_{\tau x}$ are subject to interpolation. Arrays corrected in this way are the initial data for further calculations. Direct calculation of P DFT points from P initial numbers requires approximately P^2 arithmetic operations. The number of operations can be drastically reduced by applying the FFT algorithm. This transformation is based on the factorization of the sum (3.109), i.e., on the sequential reduction of the original DFT to P_1, $P_1{\cdot}P_2$, $P_1{\cdot}P_2{\cdot}P_3$ and so on DFTs, each of which is performed on reduced arrays, and the subsequent combination of the results. Here, $P_1, P_2, \ldots P_p$ are prime factors of the number P, i.e., $P = P_1{\cdot}P_2{\cdot}\ldots{\cdot}P_p$. This procedure allows to reduce the total number of operations in calculating the DFT to the value of the order of $P\sum_{i=1}^{p} P_i$ instead of P^2. Hence, the benefit is maximal, when P is the product of the maximum number of simple factors—i.e., twos ($P = 2^p$). In this case, the FFT allows to reduce the required number of operations from P^2 to $2P{\cdot}\log_2 P$. However, the FFT hardly links the number of points in input and output arrays. The direct application of the FFT to (3.109) leads to equality of the numbers of spatial frequencies k_x, k_y to the number of spatial points, i.e., $Q = M, L = N$, and expression (3.109) takes the following form:

$$A_{x,y}(q,l) = \sum_{m}^{M-1}\sum_{n}^{N-1} E_{\tau x,y}(m,n)\cdot\exp\bigl[\mathrm{j}2\pi(q\cdot m/(M-1) + l\cdot n/(N-1))\bigr]. \tag{3.110}$$

As a result, the angular intervals of RP samples can be too large for its visual representation. In addition, when performing FFT, the angular coordinates of the sighting point (θ, φ) or (α, β) are included in the formulas for spatial frequencies in the form of the trigonometric function arguments sin and cos; therefore, to determine the antenna RP dependence directly by the angular variables, it is necessary to solve transcendental equations. Recently, the use of FFT in processing of vector $\mathbf{E}_\tau$ measured complex amplitudes on the scan plane has lost its relevance, since modern computer technology allows to perform DFT very quickly. Let us consider one of the direct DFT calculation variants (without using FFT) that allows obtaining direct dependence of the AUT RP on sighting point angular coordinates. It should be noted that in this case the minimum discrete values Δx and Δy have no fundamental limitations. On the contrary, as shown in [3], the choice of these values lesser than $\lambda/2$ leads to an increase in the accuracy of the AUT characteristics determination. After measurements, the complex amplitude can be written as a pair of amplitude and phase. Let us introduce special notation for the modules and arguments (phases) of the complex amplitudes of the vector $\mathbf{E}_\tau$ orthogonal components measured on the scan plane:

$$E_{\tau x} = |E_{\tau x}| \exp(\mathrm{j}(\arg E_{\tau x})) = |E_{\tau x}| \exp(\mathrm{j}\varphi_{\tau x}) \tag{3.111}$$

$$E_{\tau y} = \left|E_{\tau y}\right| \exp\left(\mathrm{j}\left(\arg E_{\tau y}\right)\right) = \left|E_{\tau y}\right| \exp\left(\mathrm{j}\varphi_{\tau y}\right) \tag{3.112}$$

Then, discrete analogs of expressions (3.61) and (3.62) can be written in the following form:

$$A_x(\theta,\varphi) = \sum_{m}^{M-1}\sum_{n}^{N-1} |E_{\tau x}(m,n)| \cdot \exp\left[\mathrm{j}\varphi_{\tau x}(m,n)\right] \cdot \exp\left[\mathrm{j}\psi(m,n,\theta,\varphi)\right], \tag{3.113}$$

$$A_y(\theta,\varphi) = \sum_{m}^{M-1}\sum_{n}^{N-1} \left|E_{\tau y}(m,n)\right| \cdot \exp\left[\mathrm{j}\varphi_{\tau y}(m,n)\right] \cdot \exp\left[\mathrm{j}\psi(m,n,\theta,\varphi)\right], \tag{3.114}$$

where $\Psi(m, n, \theta, \varphi) = k(m\Delta x \cos\varphi \sin\theta + n\Delta y \sin\varphi \sin\theta)$.

It is more convenient to present complex quantities $A_x(\theta, \varphi)$ and $A_y(\theta, \varphi)$ as sums of real and imaginary parts:

$$A_x(\theta,\varphi) = \mathrm{Re}\{A_x(\theta,\varphi)\} + \mathrm{jIm}\{A_x(\theta,\varphi)\} \tag{3.115}$$

$$A_y(\theta,\varphi) = \mathrm{Re}\left\{A_y(\theta,\varphi)\right\} + \mathrm{jIm}\left\{A_y(\theta,\varphi)\right\} \tag{3.116}$$

The expressions for calculating the real and imaginary parts of the complex quantities $A_x(\theta, \varphi)$ and $A_y(\theta, \varphi)$ are as follows:

$$\mathrm{Re}\{A_x(\theta,\varphi)\} = \sum_{m}^{M-1}\sum_{n}^{N-1} |E_{\tau x}(m,n)| \cdot \cos[\varphi_{\tau x}(m,n) - \psi(m,n,\theta,\varphi)], \tag{3.117}$$

$$\mathrm{Im}\{A_x(\theta,\varphi)\} = \sum_m^{M-1}\sum_n^{N-1} |E_{\tau x}(m,n)| \cdot \sin[\varphi_{\tau x}(m,n) - \psi(m,n,\theta,\varphi)], \quad (3.118)$$

$$\mathrm{Re}\{A_y(\theta,\varphi)\} = \sum_m^{M-1}\sum_n^{N-1} |E_{\tau y}(m,n)| \cdot \cos[\varphi_{\tau y}(m,n) - \psi(m,n,\theta,\varphi)], \quad (3.119)$$

$$\mathrm{Im}\{A_y(\theta,\varphi)\} = \sum_m^{M-1}\sum_n^{N-1} |E_{\tau y}(m,n)| \cdot \sin[\varphi_{\tau y}(m,n) - \psi(m,n,\theta,\varphi)]. \quad (3.120)$$

Here, θ and φ are the equidistance ($\Delta\theta = \text{const}$, $\Delta\varphi = \text{const}$) point coordinates on the sphere at which the RP should be determined. The accuracy of determining the radiation pattern, i.e., the correspondence degree of the calculated RP to its real value, depends on the measurement errors $|E_{\tau x}|$ and arg $E_{\tau x}$ on the scanning surface, the number of points at which these measurements are made, and the errors in determining their coordinates. Total antenna power RP for the spherical coordinates of sighting point is defined as follows:

$$|f(\theta,\varphi)|^2 = |f_\theta(\theta,\varphi)|^2 + |f_\varphi(\theta,\varphi)|^2 \quad (3.121)$$

Taking aforesaid into account, the expressions for partial RPs included into expression (3.121) can be represented as:

$$\begin{aligned} f_\theta(\theta,\varphi) = {} & (\mathrm{Re}\{A_x(\theta,\varphi)\} + \mathrm{jIm}\{A_x(\theta,\varphi)\})\cos\varphi \\ & + (\mathrm{Re}\{A_y(\theta,\varphi)\} + \mathrm{jIm}\{A_y(\theta,\varphi)\})\sin\varphi. \end{aligned} \quad (3.122)$$

$$\begin{aligned} f_\varphi(\theta,\varphi) = {} & -(\mathrm{Re}\{A_x(\theta,\varphi)\} + \mathrm{jIm}\{A_x(\theta,\varphi)\})\sin\varphi\cos\theta \\ & + (\mathrm{Re}\{A_y(\theta,\varphi)\} + \mathrm{jIm}\{A_y(\theta,\varphi)\})\cos\varphi\cos\theta. \end{aligned} \quad (3.123)$$

Hence,

$$\begin{aligned} |f_\theta(\theta,\varphi)|^2 = {} & \left[(\mathrm{Re}\{A_x(\theta,\varphi)\})^2 + (\mathrm{Im}\{A_x(\theta,\varphi)\})^2\right]\cos^2\varphi \\ & + \left[(\mathrm{Re}\{A_y(\theta,\varphi)\})^2 + (\mathrm{Im}\{A_y(\theta,\varphi)\})^2\right]\sin^2\varphi \\ & + \left[(\mathrm{Re}\{A_x(\theta,\varphi)\})(\mathrm{Re}\{A_y(\theta,\varphi)\}) + (\mathrm{Im}\{A_x(\theta,\varphi)\})(\mathrm{Im}\{A_y(\theta,\varphi)\})\right]\sin 2\varphi \end{aligned} \quad (3.124)$$

$$\begin{aligned} |f_\varphi(\theta,\varphi)|^2 = {} & \left[(\mathrm{Re}\{A_x(\theta,\varphi)\})^2 + (\mathrm{Im}\{A_x(\theta,\varphi)\})^2\right]\sin^2\varphi\cos^2\theta \\ & + \left[(\mathrm{Re}\{A_y(\theta,\varphi)\})^2 + (\mathrm{Im}\{A_y(\theta,\varphi)\})^2\right]\cos^2\varphi\cos^2\theta \\ & - \left[(\mathrm{Re}\{A_x(\theta,\varphi)\})(\mathrm{Re}\{A_y(\theta,\varphi)\}) + (\mathrm{Im}\{A_x(\theta,\varphi)\})(\mathrm{Im}\{A_y(\theta,\varphi)\})\right]\sin 2\varphi\cos^2\theta \end{aligned} \quad (3.125)$$

This results in the expression for total AUT power RP:

$$|f(\theta,\varphi)|^2 = \left[(\mathrm{Re}\{A_x(\theta,\varphi)\})^2 + (\mathrm{Im}\{A_x(\theta,\varphi)\})^2\right](\cos^2\varphi + \sin^2\varphi\cos^2\theta)$$
$$+\left[\left(\mathrm{Re}\{A_y(\theta,\varphi)\}\right)^2 + \left(\mathrm{Im}\{A_y(\theta,\varphi)\}\right)^2\right](\sin^2\varphi + \cos^2\varphi\cos^2\theta)$$
$$+\left[(\mathrm{Re}\{A_x(\theta,\varphi)\})\left(\mathrm{Re}\{A_y(\theta,\varphi)\}\right) + (\mathrm{Im}\{A_x(\theta,\varphi)\})\left(\mathrm{Im}\{A_y(\theta,\varphi)\}\right)\right]\sin 2\varphi\sin^2\theta \tag{3.126}$$

The total amplitude RP is a square root of expression (3.126). If the AUT is linearly polarized and vector $\mathbf{E}_\tau$ direction coincides with the direction of one of the coordinate axes (x or y), then formulas can be significantly simplified, because only one component of the PWS remains in them (A_x or A_y). For the elevation–azimuth coordinate system, the calculation expressions have the following form:

$$A_x(\alpha,\beta) = \sum_m^{M-1}\sum_n^{N-1}|E_{\tau x}(m,n)|\cdot\exp[\mathrm{j}\varphi_{\tau x}(m,n)]\cdot\exp[\mathrm{j}\psi(m,n,\alpha,\beta)] \tag{3.127}$$

$$A_y(\alpha,\beta) = \sum_m^{M-1}\sum_n^{N-1}|E_{\tau y}(m,n)|\cdot\exp[\mathrm{j}\varphi_{\tau y}(m,n)]\cdot\exp[\mathrm{j}\psi(m,n,\alpha,\beta)] \tag{3.128}$$

where $\Psi(m, n, \alpha, \beta) = k(m\Delta x\sin\alpha\cos\beta + n\Delta y\sin\beta)$.

$$A_x(\alpha,\beta) = \mathrm{Re}\{A_x(\alpha,\beta)\} + \mathrm{jIm}\{A_x(\alpha,\beta)\} \tag{3.129}$$

$$A_y(\alpha,\beta) = \mathrm{Re}\{A_y(\alpha,\beta)\} + \mathrm{jIm}\{A_y(\alpha,\beta)\} \tag{3.130}$$

$$\mathrm{Re}\{A_x(\alpha,\beta)\} = \sum_m^{M-1}\sum_n^{N-1}|E_{\tau x}(m,n)|\cdot\cos[\varphi_{\tau x}(m,n) + \psi(m,n,\alpha,\beta)] \tag{3.131}$$

$$\mathrm{Im}\{A_x(\alpha,\beta)\} = \sum_m^{M-1}\sum_n^{N-1}|E_{\tau x}(m,n)|\cdot\sin[\varphi_{\tau x}(m,n) + \psi(m,n,\alpha,\beta)] \tag{3.132}$$

$$\mathrm{Re}\{A_y(\alpha,\beta)\} = \sum_m^{M-1}\sum_n^{N-1}|E_{\tau y}(m,n)|\cdot\cos[\varphi_{\tau y}(m,n) + \psi(m,n,\alpha,\beta)] \tag{3.133}$$

$$\mathrm{Im}\{A_y(\alpha,\beta)\} = \sum_m^{M-1}\sum_n^{N-1}|E_{\tau y}(m,n)|\cdot\sin[\varphi_{\tau y}(m,n) + \psi(m,n,\alpha,\beta)] \tag{3.134}$$

$$|f(\alpha,\beta)|^2 = |f_\alpha(\alpha,\beta)|^2 + |f_\beta(\alpha,\beta)|^2 \tag{3.135}$$

$$f_\alpha(\alpha,\beta) = (\mathrm{Re}\{A_x(\alpha,\beta)\} + \mathrm{jIm}\{A_x(\alpha,\beta)\})\cos\beta$$

$$+\left(\mathrm{Re}\{A_y(\alpha,\beta)\}+\mathrm{jIm}\{A_y(\alpha,\beta)\}\right)\sin\alpha\sin\beta \tag{3.136}$$

$$f_\beta(\alpha,\beta)=\left(\mathrm{Re}\{A_y(\alpha,\beta)\}+\mathrm{jIm}\{A_y(\alpha,\beta)\}\right)\cos\alpha. \tag{3.137}$$

$$\begin{aligned}|f_\alpha(\alpha,\beta)|^2 &= \left[(\mathrm{Re}\{A_x(\alpha,\beta)\})^2+(\mathrm{Im}\{A_x(\alpha,\beta)\})^2\right]\cos^2\beta\\ &+\left[\mathrm{Re}\{A_y(\alpha,\beta)\})^2+(\mathrm{Im}\{A_y(\alpha,\beta)\})^2\right]\sin^2\alpha\cos^2\beta\\ &+2\left[\mathrm{Re}\{A_x(\alpha,\beta)\}\mathrm{Re}\{A_y(\alpha,\beta)\}+\mathrm{Im}\{A_x(\alpha,\beta)\}\mathrm{Im}\{A_y(\alpha,\beta)\}\right]\cos^2\beta\sin\alpha.\end{aligned} \tag{3.138}$$

$$\left|f_\beta(\alpha,\beta)\right|^2=\left(\mathrm{Re}\{A_y(\alpha,\beta)\}\right)^2\cos^2\alpha+\left(\mathrm{Im}\{A_y(\alpha,\beta)\}\right)^2\cos^2\alpha. \tag{3.139}$$

$$\begin{aligned}|f(\alpha,\beta)|^2 &= \left[(\mathrm{Re}\{A_x(\alpha,\beta)\})^2+(\mathrm{Im}\{A_x(\alpha,\beta)\})^2\right]\cos^2\beta\\ &+\left[\mathrm{Re}\{A_y(\alpha,\beta)\})^2+(\mathrm{Im}\{A_y(\alpha,\beta)\})^2\right](\cos^2\alpha+\sin^2\alpha\cos^2\beta)\\ &+2\left[\mathrm{Re}\{A_x(\alpha,\beta)\}\mathrm{Re}\{A_y(\alpha,\beta)\}+\mathrm{Im}\{A_x(\alpha,\beta)\}\mathrm{Im}\{A_y(\alpha,\beta)\}\right]\cos^2\beta\sin\alpha.\end{aligned} \tag{3.140}$$

The above expressions underlie the methodology for processing the measurement results used in [4].

3.5 Probe Correction

There are two methods for probe correction of planar measurements. These approaches are based on the scattering matrix theory and the Lorentz reciprocity theorem [2, 7, 8]. The first method based on scattering matrix theory is described in detail in [7].

The basic equations of the scattering theory are as follows:

$$b_0=S_{00}a_0+\int_{-\infty}^{\infty}\int_{-\infty}^{\infty}\sum_{p=1}^{2}\mathbf{S}_{0p}(\mathbf{K})\mathbf{F}_p(\mathbf{K})\mathrm{d}\mathbf{K}, \tag{3.141}$$

$$\mathbf{B}_q(\mathbf{K})=\mathbf{S}_{q0}(\mathbf{K})a_0+\int_{-\infty}^{\infty}\int_{-\infty}^{\infty}\sum_{p=1}^{2}\mathbf{S}_{qp}(\mathbf{K},\mathbf{L})\mathbf{F}_p(\mathbf{L})\mathrm{d}\mathbf{L}, \tag{3.142}$$

where indices p and q equal to 1 correspond to the front hemisphere, while indices p and q equal to 2 correspond to the back hemisphere.

a_0 is the complex amplitude of the wave propagating toward radiators (aperture) of the antenna;

b_0 is the complex amplitude of the wave propagating to the antenna load;

$\Gamma = S_{00}$ is the complex reflection coefficient of path wave from the antenna output;
$\mathbf{B}_q$ is the vector amplitude spectrum of plane waves due to radiation and scattering of the antenna;
$\mathbf{F}_p$ is the vector amplitude spectrum of plane waves incident on the antenna;
$\mathbf{S}_{q0}$ is the vector amplitude spectrum of plane waves emitted by the antenna;
$\mathbf{S}_{qp}$ is the full scattering matrix;
$\mathbf{S}_{op}$ is the vector antenna reception characteristic.

$$\mathbf{K} = k_x \mathbf{i}_x + k_y \mathbf{i}_y.$$

In Fig. 3.8, planar near-field measurement schema is presented.

As can be seen from the figure, the schema includes the AUT and a measuring probe. For the probe, as for any antenna, Eqs. (3.141) and (3.142) are also valid. All notations associated with the probe are primed. For a two-port system, the relationship between the incident, reflected, and passed waves for the AUT has the form:

$$\begin{aligned} b_0 &= S_{00}a_0 + S_{01}a_1 \\ b_1 &= S_{10}a_0 + S_{11}a_1 \end{aligned} \tag{3.143}$$

For the probe, similar relations are valid:

$$\begin{aligned} b'_0 &= S'_{00}a'_0 + S'_{02}a'_2 \\ b'_2 &= S'_{20}a'_0 + S'_{22}a'_2 \end{aligned} \tag{3.144}$$

Let us establish the relationship between the signal supplied to the AUT input—a_0, and signal received by the probe—b'_0. If Γ_l is the load reflection coefficient, then

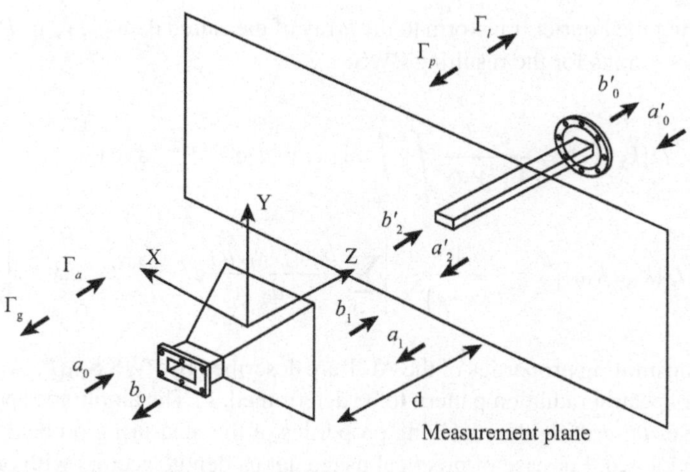

Fig. 3.8 Planar near-field measurement scheme

$a'_0 = \Gamma_l b'_0$. Let us consider the case when there are no multiple reflections between the AUT and the probe. In this case, Formula (3.142) can be simplified:

$$\mathbf{B}_1(\mathbf{K}) = \mathbf{S}_{10}(\mathbf{K})a_0 \tag{3.145}$$

Taking into account that the PWS of the plane waves incident on the probe is related to $\mathbf{B}_1(\mathbf{K})$ as $\mathbf{F}'_1(\mathbf{K}) = \mathbf{B}_1(\mathbf{K})\mathrm{e}^{-ik_x x - ik_y y - ik_z d}$, one can rewrite (3.141) as follows:

$$\begin{aligned} b'_0(x, y, d) &= S'_{00}a'_0 + \int\limits_{-\infty}^{\infty}\int\limits_{-\infty}^{\infty} \mathbf{S}'_{02}(\mathbf{K})\mathbf{F}'_1(\mathbf{K})\mathrm{d}\mathbf{K} \\ &= \Gamma_p\Gamma_l b'_0 + a_0 \int\limits_{-\infty}^{\infty}\int\limits_{-\infty}^{\infty} \mathbf{S}'_{02}(\mathbf{K})\mathbf{S}_{10}(\mathbf{K})\mathrm{e}^{-ik_x x - ik_y y - ik_z d}\mathrm{d}\mathbf{K}. \end{aligned} \tag{3.146}$$

where $\Gamma_p = S'_{00}$ is the probe reflection coefficient.

Thus, the signal received by the probe b'_0 at the point (x, y, d) is related to the input signal a_0 as follows:

$$b'_0(x, y, d) = \frac{a_0}{1 - \Gamma_p\Gamma_l} \int\limits_{-\infty}^{\infty}\int\limits_{-\infty}^{\infty} D(k_x, k_y)\mathrm{e}^{-ik_x x - ik_y y - ik_z d}\mathrm{d}k_x\mathrm{d}k_y, \tag{3.147}$$

where $D(k_x, k_y) = \mathbf{S}'_{02}(k_x, k_y)\mathbf{S}_{10}(k_x, k_y)$ is the resulting angular spectrum and is defined by the scalar product of the antenna transmitting characteristic and the probe receiving characteristic.

Let us denote impedance mismatch coefficient between the probe and the port of load connected to it as $\gamma = \frac{1}{1-\Gamma_l\Gamma_p}$.

Applying the Fourier transform to the array of measured data $b'_0(x, y, d)$, one can obtain expressions for the resulting PWS:

$$D(k_x, k_y) = \frac{\mathrm{e}^{ik_z d}}{4\pi^2\gamma a_0} \int\limits_{-\infty}^{\infty}\int\limits_{-\infty}^{\infty} b'_0(x, y, d)\mathrm{e}^{ik_x x + ik_y y}\mathrm{d}x\mathrm{d}y \tag{3.148}$$

$$\left|D(k_{x0}, k_{y0})\right|^2 = \frac{1}{(4\pi^2\gamma)^2} \cdot \left|\sum_{x_i, y_i} \frac{b'_0(x_i, y_i, d)}{a_0}\mathrm{e}^{ik_{x0}x_i + ik_{y0}y_i}\Delta x\Delta y\right|^2$$

The transmitting properties of the AUT are described by PWS $\mathbf{S}_{10}(k_x, k_y)$. This is the vector spectral radiation pattern to be determined, P. The amplitude spectrum of plane waves describing the receiving properties of the measuring probe $\mathbf{S}'_{02}(k_x, k_y)$ should be known. For ease of practical usage, let us denote vectors with only transverse (to the propagation direction z) components with capital letter $\mathbf{S}_{10}$ and vectors including also axial component with small letter $\mathbf{s}_{10}$ [7]. The relationships of these vectors with the resulting angular spectrum are identical:

$$D(k_x, k_y) = \mathbf{S}'_{02}(k_x, k_y)\mathbf{S}_{10}(k_x, k_y) = \mathbf{s}'_{02}(k_x, k_y)\mathbf{s}_{10}(k_x, k_y). \tag{3.149}$$

Let us find vector spectral radiation pattern of the AUT. In the simplest case, when antenna is linearly polarized and the level of the cross-polarization component is small, then to find the desired PWS of the AUT is enough to measure only one polarization. In the chosen coordinate system, this is the X or Y component of the field on the scan plane. In this case, taking into account (3.149) expression (3.148) takes a simplified form:

$$s_{10}(k_x, k_y) = \frac{a_0}{s'_{02}(k_x, k_y)} \frac{e^{ik_z d}}{4\pi^2 \gamma} \int_{-\infty}^{\infty}\int_{-\infty}^{\infty} b'_0(x, y, d) e^{ik_x x + ik_y y} dxdy. \tag{3.150}$$

From this expression, one can immediately find the desired component of the spectral radiation pattern of the AUT, assuming that the same component of the spectral diagram for the probe $s'_{02}(k_x, k_y)$ is known. In the general case, in order to find the main and cross-polarization components, two measurements are required for two orthogonal polarizations. Then, to find all field components, it is necessary to solve the system of equations, which is considered below. If $b'_0(x, y, d)$ and $b''_0(x, y, d)$ are the complex amplitudes of the probe output signals during measurements on two orthogonal polarizations at a point (x, y) on the scan plane, then the system of equations can be written in the following form:

$$\begin{pmatrix} s'_{02m}(k_x, k_y) & s'_{02c}(k_x, k_y) \\ s''_{02m}(k_x, k_y) & s''_{02c}(k_x, k_y) \end{pmatrix} \cdot \begin{pmatrix} s_{10m}(k_x, k_y) \\ s_{10c}(k_x, k_y) \end{pmatrix} = \begin{pmatrix} D'(k_x, k_y) \\ D''(k_x, k_y) \end{pmatrix}, \tag{3.151}$$

where

$$D'(k_x, k_y) = \frac{e^{ik_z d}}{4\pi^2 \gamma' a_0} \int_{-\infty}^{\infty}\int_{-\infty}^{\infty} b'_0(x, y, d) e^{ik_x x + ik_y y} dxdy, \quad \gamma' = \frac{1}{1 - \Gamma_l \Gamma'_p}, \tag{3.152}$$

$$D''(k_x, k_y) = \frac{e^{ik_z d}}{4\pi^2 \gamma'' a_0} \int_{-\infty}^{\infty}\int_{-\infty}^{\infty} b''_0(x, y, d) e^{ik_x x + ik_y y} dxdy, \quad \gamma'' = \frac{1}{1 - \Gamma_l \Gamma''_p}. \tag{3.153}$$

Probe receiving characteristics for two orthogonal polarizations are characterized by PWSs $\mathbf{s}'_{02}(k_x, k_y)$ and $\mathbf{s}''_{02}(k_x, k_y)$ accordingly. Index "m" corresponds to the main component and primed measurement results, and index "c"—to the cross-component and double primed measurement results. The subscripts "m" and "c" can correspond to θ or φ components in spherical or α or β ones in azimuth–elevation system. This representation is correct, because in these coordinate systems, axial component E_r at far-field region is negligible. If PWSs of the probe are defined in a Cartesian coordinate system, then all three components $s'_{02x}(k_x, k_y)$, $s'_{02y}(k_x, k_y)$, $s'_{02z}(k_x, k_y)$ should be taken into account. In this case, $s'_{02z}(k_x, k_y)$ is expressed as linear combination of $s'_{02x}(k_x, k_y)$ and $s'_{02y}(k_x, k_y)$ based on the div $\mathbf{E} = 0$ condition written in the form $\mathbf{k} \cdot \mathbf{S}'_{02} = 0$. Similarly, $s''_{02z}(k_x, k_y)$ is expressed as a linear combination of $s''_{02x}(k_x, k_y)$ and $s''_{02y}(k_x, k_y)$. The probe correction in the case when the components of its field are expressed in a Cartesian coordinate system is examined in detail in [9]. In practice, the representation of the probe field components in a Cartesian system is not convenient and is rarely used, so representation in spherical coordinate system will be considered further.

Solving system of Eq. (3.151), one can obtain:

$$s_{10m}(k_x, k_y) = \frac{D'(k_x, k_y) s''_{02c}(k_x, k_y) - D''(k_x, k_y) s'_{02c}(k_x, k_y)}{s'_{02m}(k_x, k_y) s''_{02c}(k_x, k_y) - s''_{02m}(k_x, k_y) s'_{02c}(k_x, k_y)} \tag{3.154}$$

$$s_{10c}(k_x, k_y) = \frac{s'_{02m}(k_x, k_y) D''(k_x, k_y) - s''_{02m}(k_x, k_y) D'(k_x, k_y)}{s'_{02m}(k_x, k_y) s''_{02c}(k_x, k_y) - s''_{02m}(k_x, k_y) s'_{02c}(k_x, k_y)} \tag{3.155}$$

According to the reciprocity theorem, the relationship between the antenna transmitting and receiving spectral characteristics has the following form [10]:

$$\mathbf{S}_{02}(\mathbf{K}) = -\frac{\eta}{\eta_0} \mathbf{S}_{10}(-\mathbf{K}), \tag{3.156}$$

$$\mathbf{s}_{02}(\mathbf{K}) = \frac{\eta}{\eta_0} \frac{k_z}{k} \mathbf{s}_{10}(-\mathbf{K}), \tag{3.157}$$

where $\eta = \sqrt{\varepsilon/\mu}$ is wave conductivity of the medium.

η_0 is the antenna conductivity.

Let us analyze the results of probe correction based on Lorentz reciprocity theorem. It is known by practice that the product of antenna PWS and probe PWS is equal [2, 10]:

$$\mathbf{A}(\mathbf{K})\mathbf{G}(-\mathbf{K}) = \left(\frac{1}{2\pi}\right)^2 \frac{1}{8\pi^2} \frac{\omega\mu}{k_z} \int_{-\infty}^{\infty} \int_{-\infty}^{\infty} P_b(x, y, d) e^{i\mathbf{k}\mathbf{r}} dx dy, \tag{3.158}$$

where

$$P_b(x, y, d) = 8\pi^2\eta_0\left(1 - \Gamma_l\Gamma'_p\right)\frac{b'_0(x, y, d)}{a_0}. \tag{3.159}$$

Using expression (3.157), expression (3.158) can be reduced to (3.148), (3.149) derived with the scattering matrix theory. Let us write the relationship of the antenna PWS with its RP and relation of the probe PWS with its RP:

$$\mathbf{E}(\mathbf{r}) = i\frac{2\pi}{r}k_z\mathbf{A}(\mathbf{K})e^{-i\mathbf{kr}}, \quad \mathbf{E}'_p(\mathbf{r}') = i\frac{2\pi}{r}k_z\mathbf{G}(\mathbf{K}')e^{-i\mathbf{kr}}. \tag{3.160}$$

Now let us find relationship between antenna and probe vector RPs using (3.158) and (3.160) (omitting $i\frac{e^{-i\mathbf{kr}}}{r}$ factor)

$$\mathbf{E}(\mathbf{r})\mathbf{E}'_p(\mathbf{r}') = k\frac{\eta_0}{\eta}\frac{1}{\left(1 - \Gamma_l\Gamma'_p\right)a_0}k_z\int\limits_{-\infty}^{\infty}\int\limits_{-\infty}^{\infty}b'_0(x, y, d)e^{i\mathbf{kr}}dxdy \tag{3.161}$$

Now let us rewrite the expression (3.161) for two most common coordinate systems: azimuth–elevation and spherical one. It should be noted that in the coordinate systems considered above, the E_r component of the vector radiation pattern in the far-field region is negligible. First, consider the azimuth–elevation coordinate system. In Fig. 3.9, relative position of the antenna coordinate system and the probe coordinate system is shown.

The unit vectors of these coordinate systems are related by the following expressions:

$$\mathbf{i}_x = -\mathbf{i}'_x, \quad \mathbf{i}_y = \mathbf{i}'_y, \quad \mathbf{i}_z = -\mathbf{i}'_z. \tag{3.162}$$

wherein $\mathbf{k}' = k'_x\mathbf{i}'_x + k'_y\mathbf{i}'_y + k'_z\mathbf{i}'_z = k_x(-\mathbf{i}_x) + (-k_y)\mathbf{i}_y + k_z(-\mathbf{i}_z) = -\mathbf{k}$.

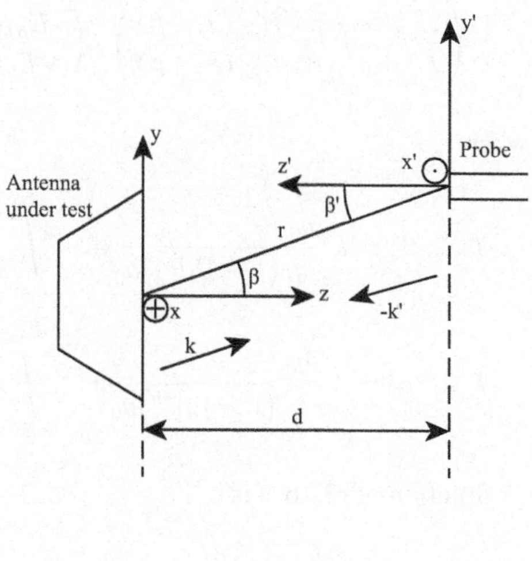

Fig. 3.9 Relative position of the antenna coordinate system and probe coordinate system in the azimuth–elevation coordinate system

The azimuth–elevation coordinate system unit vectors are related to the Cartesian coordinate system unit vectors with known equations [1, 11]:

$$\begin{pmatrix} \mathbf{i}_R \\ \mathbf{i}_\alpha \\ \mathbf{i}_\beta \end{pmatrix} = \begin{pmatrix} \cos\beta\sin\alpha & \sin\beta & \cos\beta\cos\alpha \\ \cos\alpha & 0 & -\sin\alpha \\ -\sin\alpha\sin\beta & \cos\beta & -\cos\alpha\sin\beta \end{pmatrix} \cdot \begin{pmatrix} \mathbf{i}_x \\ \mathbf{i}_y \\ \mathbf{i}_z \end{pmatrix} \tag{3.163}$$

wherein

$$\begin{cases} k_x = k\cos\beta\sin\alpha \\ k_y = k\sin\beta \\ k_z = k\cos\beta\cos\alpha \end{cases} \tag{3.164}$$

$$\mathbf{E}(\mathbf{r})\mathbf{E}'_p(\mathbf{r}') = E_\alpha(\alpha,\beta)E'_{p\alpha}(\alpha',\beta')\mathbf{i}_\alpha\mathbf{i}'_\alpha + E_\beta(\alpha,\beta)E'_{p\beta}(\alpha',\beta')\mathbf{i}_\beta\mathbf{i}'_\beta \tag{3.165}$$

Taking into account (3.162) and (3.163) for the coordinate systems of the antenna and the probe, one can note that $\mathbf{i}_\alpha\mathbf{i}'_\alpha = -1$, and $\mathbf{i}_\beta\mathbf{i}'_\beta = 1$, where $\alpha' = \alpha$ and $\beta' = -\beta$. Thus, expression (3.161) in the azimuth–elevation system has the form:

$$\begin{aligned} &E_\beta(\alpha,\beta)E'_{p\beta}(\alpha,-\beta) - E_\alpha(\alpha,\beta)E'_{p\alpha}(\alpha,-\beta) \\ &= k\frac{\eta_0}{\eta}\frac{1}{\left(1-\Gamma_l\Gamma'_p\right)a_0}k_z\int_{-\infty}^{\infty}\int_{-\infty}^{\infty} b'_0(x,y,d)\mathrm{e}^{i\mathbf{kr}}\mathrm{d}x\mathrm{d}y \end{aligned} \tag{3.166}$$

Similarly to finding the spectral radiation pattern components, it is required to carry out two near-field measurements of two orthogonal polarizations to find $E_\alpha(\alpha,\beta)$ and $E_\beta(\alpha,\beta)$. In this case, if the probe RP is known for both polarizations ($\mathbf{E}'_\mathbf{p}$ and $\mathbf{E}''_\mathbf{p}$), then the components of the antenna RP can be found from the system of equations similar to (3.151)

$$\begin{pmatrix} E'_{p\beta}(\alpha,-\beta) & E'_{p\alpha}(\alpha,-\beta) \\ E''_{p\beta}(\alpha,-\beta) & E''_{p\alpha}(\alpha,-\beta) \end{pmatrix} \cdot \begin{pmatrix} E_\beta(\alpha,\beta) \\ -E_\alpha(\alpha,\beta) \end{pmatrix} = \frac{k_z}{k}\begin{pmatrix} I'(\alpha,\beta) \\ I''(\alpha,\beta) \end{pmatrix}, \tag{3.167}$$

where

$$I'(\alpha,\beta) = k^2\frac{\eta_0}{\eta}\frac{1}{\left(1-\Gamma_l\Gamma'_p\right)a_0}\mathrm{e}^{ik_zd}\int_{-\infty}^{\infty}\int_{-\infty}^{\infty} b'_0(x,y,d)\mathrm{e}^{ik_xx+ik_yy}\mathrm{d}x\mathrm{d}y, \tag{3.168}$$

$$I''(\alpha,\beta) = k^2\frac{\eta_0}{\eta}\frac{1}{\left(1-\Gamma_l\Gamma''_p\right)a_0}\mathrm{e}^{ik_zd}\int_{-\infty}^{\infty}\int_{-\infty}^{\infty} b''_0(x,y,d)\mathrm{e}^{ik_xx+ik_yy}\mathrm{d}x\mathrm{d}y. \tag{3.169}$$

Solution of (3.167) is

$$E_\beta(\alpha,\beta) = \frac{\cos\beta\cos\alpha}{\Delta(\alpha,\beta)}\left[I'(\alpha,\beta)E''_{p\alpha}(\alpha,-\beta) - I''(\alpha,\beta)E'_{p\alpha}(\alpha,-\beta)\right], \quad (3.170)$$

$$E_\alpha(\alpha,\beta) = \frac{\cos\beta\cos\alpha}{\Delta(\alpha,\beta)}\left[I'(\alpha,\beta)E'_{p\beta}(\alpha,-\beta) - I''(\alpha,\beta)E''_{p\beta}(\alpha,-\beta)\right], \quad (3.171)$$

where

$$\Delta(\alpha,\beta) = E'_{p\beta}(\alpha,-\beta)E''_{p\alpha}(\alpha,-\beta) - E''_{p\beta}(\alpha,-\beta)E'_{p\alpha}(\alpha,-\beta). \quad (3.172)$$

Consider spherical coordinate system. In Fig. 3.10, relative position of the antenna coordinate system and the probe coordinate system is shown.

The relationship of the unit vectors of the spherical coordinate system and Cartesian one is described by the following well-known matrix [1, 11]:

$$\begin{pmatrix} \mathbf{i}_R \\ \mathbf{i}_\theta \\ \mathbf{i}_\varphi \end{pmatrix} = \begin{pmatrix} \sin\vartheta\cos\varphi & \sin\vartheta\sin\varphi & \cos\vartheta \\ \cos\vartheta\cos\varphi & \cos\vartheta\sin\varphi & -\sin\vartheta \\ -\sin\varphi & \cos\varphi & 0 \end{pmatrix} \cdot \begin{pmatrix} \mathbf{i}_x \\ \mathbf{i}_y \\ \mathbf{i}_z \end{pmatrix} \quad (3.173)$$

wherein

$$\begin{cases} k_x = k\sin\theta\cos\varphi \\ k_y = k\sin\theta\sin\varphi \\ k_z = k\cos\theta \end{cases} \quad (3.174)$$

Similar to (3.165), one can write:

$$\mathbf{E}(\mathbf{r})\mathbf{E}'_p(\mathbf{r}') = E_\theta(\theta,\varphi)E'_{p\theta}(\theta',\varphi')\mathbf{i}_\theta\mathbf{i}'_\theta + E_\varphi(\theta,\varphi)E'_{p\varphi}(\theta',\varphi')\mathbf{i}_\varphi\mathbf{i}'_\varphi. \quad (3.175)$$

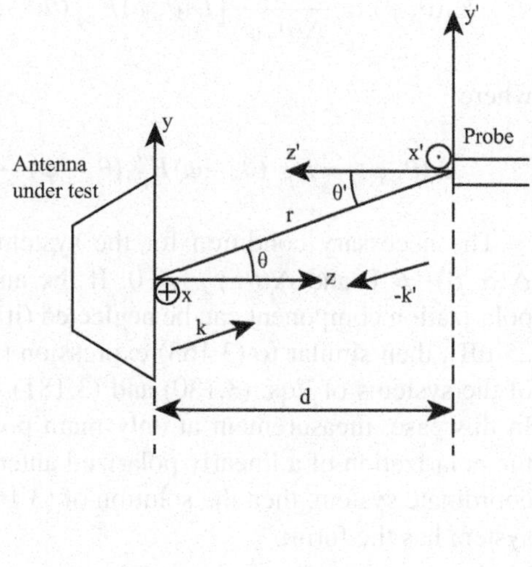

Fig. 3.10 Relative position of the antenna coordinate system and the probe coordinate system in the spherical coordinate system

Using (3.149) and (3.157), one can find that $\mathbf{i}_\theta \mathbf{i}'_\theta = -1, \mathbf{i}_\varphi \mathbf{i}'_\varphi = 1$, wherein $\theta' = \theta, \varphi' = -\varphi$. Thus, expression (3.161) in the spherical coordinate system has the form:

$$E_\varphi(\theta,\varphi)E'_{p\varphi}(\theta,-\varphi) - E_\theta(\theta,\varphi)E'_{p\theta}(\theta,-\varphi)$$
$$= k\frac{\eta_0}{\eta}\frac{1}{\left(1-\Gamma_l\Gamma'_p\right)a_0}k_z\int\limits_{-\infty}^{\infty}\int\limits_{-\infty}^{\infty} b'_0(x,y,d)\mathrm{e}^{i\mathbf{kr}}\mathrm{d}x\mathrm{d}y \quad (3.176)$$

For two measurements of the orthogonal polarizations, system gets the form:

$$\begin{pmatrix} E'_{p\varphi}(\theta,-\varphi) & E'_{p\theta}(\theta,-\varphi) \\ E''_{p\varphi}(\theta,-\varphi) & E''_{p\theta}(\theta,-\varphi) \end{pmatrix} \cdot \begin{pmatrix} E_\varphi(\theta,\varphi) \\ -E_\theta(\theta,\varphi) \end{pmatrix} = \frac{k_z}{k}\begin{pmatrix} I'(\theta,\varphi) \\ I''(\theta,\varphi) \end{pmatrix}, \quad (3.177)$$

where

$$I'(\theta,\varphi) = k^2\frac{\eta_0}{\eta}\frac{1}{\left(1-\Gamma_l\Gamma'_p\right)a_0}\mathrm{e}^{ik_z d}\int\limits_{-\infty}^{\infty}\int\limits_{-\infty}^{\infty} b'_0(x,y,d)\mathrm{e}^{ik_x x+ik_y y}\mathrm{d}x\mathrm{d}y, \quad (3.178)$$

$$I''(\theta,\varphi) = k^2\frac{\eta_0}{\eta}\frac{1}{\left(1-\Gamma_l\Gamma''_p\right)a_0}\mathrm{e}^{ik_z d}\int\limits_{-\infty}^{\infty}\int\limits_{-\infty}^{\infty} b''_0(x,y,d)\mathrm{e}^{ik_x x+ik_y y}\mathrm{d}x\mathrm{d}y. \quad (3.179)$$

The solution of Eq. (3.177) can be obtained in the following form:

$$E_\varphi(\theta,\varphi) = \frac{\cos\theta}{\Delta(\theta,\varphi)}\left[I'(\theta,\varphi)E''_{p\theta}(\theta,-\varphi) - I''(\theta,\varphi)E'_{p\theta}(\theta,-\varphi)\right], \quad (3.180)$$

$$E_\theta(\theta,\varphi) = \frac{\cos\theta}{\Delta(\theta,\varphi)}\left[I'(\theta,\varphi)E'_{p\varphi}(\theta,-\varphi) - I''(\theta,\varphi)E''_{p\varphi}(\theta,-\varphi)\right], \quad (3.181)$$

where

$$\Delta(\theta,\varphi) = E'_{p\varphi}(\theta,-\varphi)E''_{p\theta}(\theta,-\varphi) - E''_{p\varphi}(\theta,-\varphi)E'_{p\theta}(\theta,-\varphi). \quad (3.182)$$

The necessary condition for the systems (3.180), (3.181) to have solution is $\Delta(\alpha,\beta) \neq 0$ and $\Delta(\theta,\varphi) \neq 0$. If the antenna is linearly polarized, and cross-polarization component can be neglected (it is less than the main one by at least 20–25 dB), then similar to (3.163) expression (3.174) can be simplified. Thus, instead of the systems of Eqs. (3.180) and (3.181), only one equation can be left to solve. In this case, measurement at only main polarization is sufficient. For example, if the polarization of a linearly polarized antenna is parallel to the *OX*-axis of chosen coordinate system, then the solution of (3.160) in the azimuth–elevation coordinate system has the form:

$$E_\alpha(\alpha,\beta) = \frac{k_z I'_x(\alpha,\beta)}{E'_{p\alpha}(\alpha,-\beta)} \tag{3.183}$$

where

$$I'_x(\alpha,\beta) = k^2 \frac{\eta_0}{\eta} \frac{1}{(1-\Gamma_l \Gamma'_p)a_0} \mathrm{e}^{ik_z d} \int\limits_{-\infty}^{\infty}\int\limits_{-\infty}^{\infty} b'_{0x}(x,y,d)\mathrm{e}^{ik_x x + ik_y y}\mathrm{d}x\mathrm{d}y \tag{3.184}$$

If the polarization of the antenna is parallel to the OY-axis,

$$E_\beta(\alpha,\beta) = \frac{k_z I'_y(\alpha,\beta)}{E'_{p\beta}(\alpha,-\beta)} \tag{3.185}$$

where

$$I'_y(\alpha,\beta) = k^2 \frac{\eta_0}{\eta} \frac{1}{(1-\Gamma_l \Gamma'_p)a_0} \mathrm{e}^{ik_z d} \int\limits_{-\infty}^{\infty}\int\limits_{-\infty}^{\infty} b'_{0y}(x,y,d)\mathrm{e}^{ik_x x + ik_y y}\mathrm{d}x\mathrm{d}y. \tag{3.186}$$

In Fig. 3.11, RP main component of the etalon horn antenna in azimuth–elevation coordinate system without probe correction and with the probe correction is shown. As follows from the graphs, the difference between the calculation results increases with the angle distance from the zero direction. The constant $k^2 \frac{\eta_0}{\eta} \frac{1}{(1-\Gamma_l \Gamma'_p)a_0}$ affects only the absolute value of the RP. Let us consider some practical aspects of

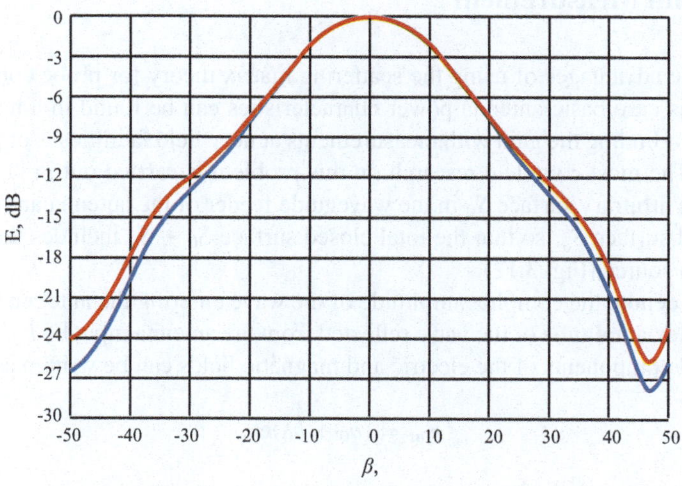

Fig. 3.11 RP main component of the etalon horn antenna in the azimuth–elevation coordinate system without probe correction (blue line) and with the one (red line)

experimental RPs calculating with probe correction. From the expressions (3.168)–(3.172) and (3.178)–(3.182), it follows that due to finite scan area and the fact that the scan steps along the X and Y coordinates do not tend to zero, all integrals in $I'(\alpha, \beta)$, $I''(\alpha, \beta)$, $I'(\theta, \varphi)$ and $I''(\theta, \varphi)$ become finite sums. In this regard, during the measurements, it is necessary to select the scan region and the scan step so as to ensure the minimum error associated with the transition from the integral to the finite integral sum. The scan step should not exceed $\lambda/2$ that is necessary to fulfill the conditions of the sampling theorem, which describes the exact reconstruction of a continuous spectrally bounded function from a set of discrete samples [1]. In practice, it is desirable to use a smaller step, taking into account that the vector network analyzer has time to sweep the frequency at each point. Otherwise, the velocity of the probe will have to be reduced. For the same scan area, a smaller step will provide a larger number of points and, therefore, a better averaging of random noise. Scan area dimensions should be chosen individually based on properties of each measured antenna. As a rule, for narrow-beam antennas, it is advisable to choose the boundaries of the scan region in such a way as to achieve a decrease in the amplitude distribution at the edges at a level of about −30 dB relative to the maximum value of the field amplitude on the scan plane. If, however, a noticeable field discontinuity appears at the edge of the scan region, then there is a probability of the Gibbs effect occurrence when calculating the AUT RP from the amplitude–phase distribution on the scan plane [12]. As a result, the calculated antenna RP can have oscillations. It is advisable to choose the distance from the AUT no more than 3λ, because, as it pursues from the foregoing, bigger distance requires bigger scan area.

3.6 Gain Measurement

One of the advantages of using the scattering matrix theory for probe correction is that in this case basic antenna power characteristics can be found in a handy way. The task of finding the gain with measurements at near-field facilities is of particular interest. The most complete research on this problem is carried out in [13]. Let us choose an arbitrary surface S_0 in the waveguide feeder of the antenna and define an additional surface S_a, so that the total closed surface $S_0 + S_a$ includes the antenna excitation source (Fig. 3.12).

Let us denote the complex amplitude of the wave entering the antenna by a_0 and the complex amplitude of the wave reflected from the antenna input by b_0. Then, the tangential components of the electric and magnetic fields can be written as follows:

$$\mathbf{E_{ot}} = (a_0 + b_0)\mathbf{e_o}, \tag{3.187}$$

$$\mathbf{H_{ot}} = \eta_0(a_0 - b_0)\mathbf{h_o}, \tag{3.188}$$

Fig. 3.12 Transmitting antenna scheme

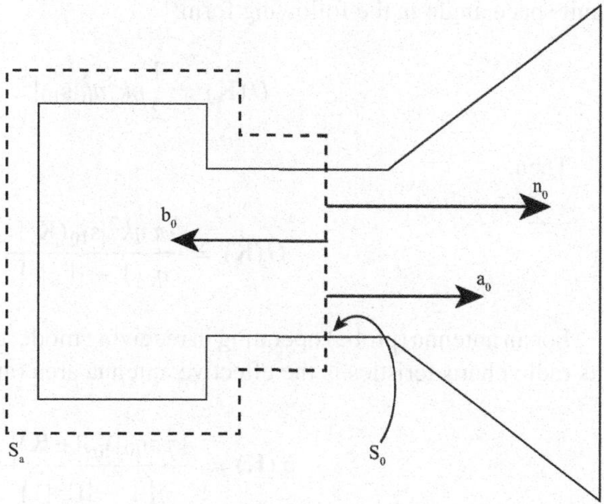

where $\mathbf{e_o}$ and $\mathbf{h_o}$ are unit vectors of electric and magnetic fields on the surface S_0 which satisfy the following condition:

$$\mathbf{h_0} = \eta_0\eta[\mathbf{N} \times \mathbf{e_o}], \quad \int_{S_0} [\mathbf{e_o} \times \mathbf{h_0}]\mathbf{N}\mathrm{d}S = 1. \tag{3.189}$$

The total power passing through the surface S_0, in this case is determined as:

$$P_0 = \frac{1}{2}\mathrm{Re}\int_{S_0} \left[E_{\mathrm{ot}} \times H^*_{\mathrm{ot}}\right]N\mathrm{d}S = \frac{1}{2}\eta_0\left(|a_0|^2 - |b_0|^2\right) = \frac{1}{2}\eta_0|a_0|^2\left(1 - |\Gamma_a|^2\right), \tag{3.190}$$

where $\Gamma_a = b_0/a_0$ is the antenna input reflection coefficient.

The antenna gain (in relative units) is defined as:

$$G(\mathbf{K}) = \frac{4\pi U(\mathbf{K})}{P_0}, \tag{3.191}$$

where P_0 is the power supplied to the input of the antenna.

U ($\mathbf{K}$) is the power radiated per unit space angle in the far-field region in direction determined by $\mathbf{K}$.

The power radiated per unit space angle in far-field region is expressed as $U(\mathbf{K}) = \frac{r^2}{2}\eta|\mathbf{E}(\mathbf{r})|^2$, where $\eta = \sqrt{\varepsilon/\mu}$ is the characteristic conductivity of the medium [9]. Using the asymptotic representation of the far field, let us write power radiated per unit space angle in the following form:

$$U(\mathbf{K}) = \frac{1}{2}\eta k_z^2 a_0^2 |\mathbf{s}_{10}|^2. \tag{3.192}$$

Then,

$$G(\mathbf{K}) = \frac{4\pi \eta k_z^2 |\mathbf{s}_{10}(\mathbf{K})|^2}{\eta_0 \left(1 - |\Gamma_a|^2\right)}. \tag{3.193}$$

For an antenna (probe) operating in receiving mode, the main parameter describing its radio characteristics is the effective antenna area (aperture) [7]:

$$\sigma(\mathbf{K}) = \frac{4\pi^2 \eta_0 \left|s'_{02}(-\mathbf{K})\right|^2}{\eta \left(1 - \left|\Gamma'_p\right|^2\right)}. \tag{3.194}$$

The gain and effective area of the reciprocal antenna are related by a known expression [1, 7]:

$$\sigma(\mathbf{K}) = \frac{\lambda^2}{4\pi} G(-\mathbf{K}). \tag{3.195}$$

Let us find the square of the expression (3.149) absolute value. Taking into account that (3.193), (3.194), and (3.195) should be determined at the point $\left(k_{x0}, k_{y0}\right)$ corresponding to the direction of the main antenna lobe, one can obtain:

$$\left|D\left(k_{x0}, k_{y0}\right)\right|^2 = \frac{\left(1 - |\Gamma_a|^2\right) G_a\left(k_{x0}, k_{y0}\right)}{4\pi k^2} \cdot \frac{\left(1 - \left|\Gamma'_p\right|^2\right) G_p\left(k_{x0}, k_{y0}\right)}{4\pi k^2} \tag{3.196}$$

Using expressions (3.148) and (3.151), let us find the formula for antenna gain calculation:

$$G_a\left(k_{x0}, k_{y0}\right) = \frac{1}{G_p\left(k_{x0}, k_{y0}\right)} \left(\frac{4\pi}{\lambda^2}\right)^2 \cdot \frac{\left|1 - \Gamma_l \Gamma'_p\right|^2}{\left(1 - \left|\Gamma'_p\right|^2\right)\left(1 - |\Gamma_a|^2\right)} \cdot \left| \sum_{x_i, y_i} \frac{b'_0(x_i, y_i, d)}{a_0} \mathrm{e}^{ik_{x0}x_i + ik_{y0}y_i} \Delta x \Delta y \right|^2 \tag{3.197}$$

In practice, to determine the input signal a_0, the transmission coefficient of entire cable system (which consists of antenna cable and probe cable) should be measured. To do this, it is necessary to connect cables to each other and measure the transmission coefficient value of this system. Connected cables form transmission line with two inhomogeneities, for which the following relationship between the measured transmission coefficient of the entire cable system a_n and the input signal a_0 is valid [14]:

$$a_0 = \frac{1 - \Gamma_g \Gamma_l}{1 - \Gamma_g \Gamma_a} a_n, \tag{3.198}$$

where Γ_g is the generator reflection coefficient.

Then, the expression for gain calculation takes the final form:

$$\mathbf{G_a}\left(\mathbf{k_{x0}}, \mathbf{k_{y0}}\right) = \left(\frac{4\pi}{\lambda^2}\right)^2 \frac{\mathbf{M}}{\mathbf{G_p}\left(\mathbf{k_{x0}}, \mathbf{k_{y0}}\right)} \left| \sum_{\mathbf{x_m}, \mathbf{y_m}} \frac{\mathbf{b}_0'(\mathbf{x_m}, \mathbf{y_m}, \mathbf{d})}{\mathbf{a_n}} \mathbf{e}^{\mathbf{i}\mathbf{k}_{x0}\mathbf{x_m} + \mathbf{i}\mathbf{k}_{y0}\mathbf{y_m}} \Delta\mathbf{x}\Delta\mathbf{y} \right|^2, \tag{3.199}$$

where $M = \frac{\left|1-\Gamma_l \Gamma_p'\right|^2 \left|1-\Gamma_g \Gamma_a\right|^2}{\left(1-\left|\Gamma_p'\right|^2\right)\left(1-\left|\Gamma_a\right|^2\right)\left|1-\Gamma_g \Gamma_l\right|^2}$ is the system mismatch coefficient.

Expression (3.197) can also be written for realized gains of the AUT and measuring probe, which differ from the gains by the amount of loss due to path (feeder) mismatch:

$$\mathbf{G_{ra}}\left(\mathbf{k_{x0}}, \mathbf{k_{y0}}\right) = \left(\frac{4\pi}{\lambda^2}\right)^2 \frac{\mathbf{M_r}}{\mathbf{G_{rp}}\left(\mathbf{k_{x0}}, \mathbf{k_{y0}}\right)} \left| \sum_{\mathbf{x_m}, \mathbf{y_m}} \frac{\mathbf{b}_0'(\mathbf{x_m}, \mathbf{y_m}, \mathbf{d})}{\mathbf{a_n}} \mathbf{e}^{\mathbf{i}\mathbf{k}_{x0}\mathbf{x_m} + \mathbf{i}\mathbf{k}_{y0}\mathbf{y_m}} \Delta\mathbf{x}\Delta\mathbf{y} \right|^2 \tag{3.200}$$

where $M_r = \frac{\left|1-\Gamma_l \Gamma_p'\right|^2 \left|1-\Gamma_g \Gamma_a\right|^2}{\left|1-\Gamma_g \Gamma_l\right|^2}$.

In Fig. 3.13, dependence of the mismatch coefficient M on the voltage standing wave ratio coefficient (VSWR) is shown.

Now let us discuss some practical questions regarding the near-field gain measurement. To determine the gain accurately, it is necessary to intercept all the energy emitted by the AUT. In this regard, firstly, it is required to select correctly the scan steps in vertical and horizontal coordinates, and secondly, to select accurately the size of the scan area. Step and scan area are selected according to the same recommendations as in the previous section. If one selects the scan region in such a way as to reduce the amplitude of the AUT field at its edges to −30 dB relative to the maximum value of the field amplitude on the scan plane, then the error in determining the absolute value of the RP maximum caused by the scan region truncation does not exceed 0.1 dB. Choosing the boundaries of the region wider, as a rule, does not make sense, because instead of receiving additional information, only the accumulation of noise will occur. If the scan region is selected in such a way that the decrease in the

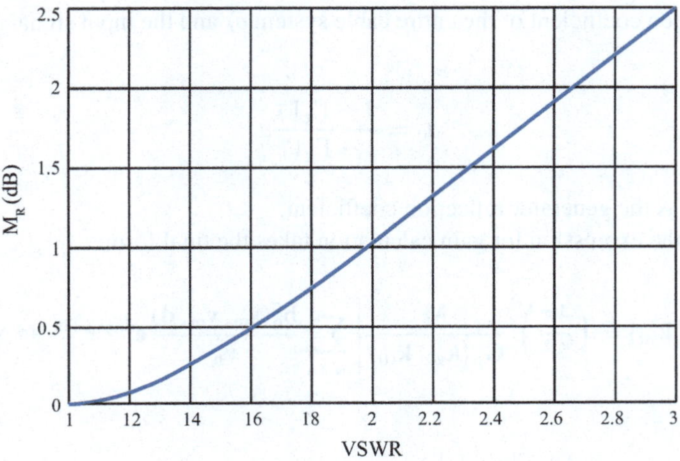

Fig. 3.13 Dependence of the mismatch coefficient *M* on the VSWR

amplitude of the signal at its edge is (−20 … −25) dB, then in this case the error caused by the truncation of the scan region increases and can reach 0.4 dB.

There are three basic methods to find the gain with near-field measurements:

- Direct method;
- Comparison with etalon antenna method;
- Three-antenna method.

Direct method. The AUT gain can be found directly from the results of measuring the amplitude–phase distribution on the scan plane if the following values are known: probe gain, the transmission coefficient of the entire cable system of the facility a_n, and the reflection coefficients from all inhomogeneities in the path through which the signal received by the probe passes (i.e., Γ_g, Γ_l, Γ_a and Γ'_p). In this case, the gain of the AUT is calculated by the expressions (3.199) or (3.200).

Comparison method. To find the gain of the AUT with comparison method, the presence of an etalon (standard) antenna with a known gain is required. In this case, two measurements of amplitude–phase distribution on the scan plane are carried out under the same conditions, first for the AUT and then for the standard antenna. The strict gain calculation with comparison method is carried out with the expression:

$$G_a\left(k_{x0}, k_{y0}\right) = \frac{\left|\sum\limits_{x_m, y_n} b'_{0a}(x_m, y_n)\mathrm{e}^{ik_{x0}x_m + ik_{y0}y_n}\right|^2}{\left|\sum\limits_{x_m, y_n} b'_{0\mathrm{ST}}(x_m, y_n)\mathrm{e}^{ik_{x0}x_m + ik_{y0}y_n}\right|^2} \frac{\left|1 - \Gamma_g\Gamma_a\right|^2}{\left|1 - \Gamma_g\Gamma_{\mathrm{ST}}\right|^2}$$

$$\frac{\left(1-|\Gamma_{\mathrm{ST}}|^2\right)}{\left(1-|\Gamma_a|^2\right)} G_{\mathrm{ST}}\left(k_{x0}, k_{y0}\right), \tag{3.201}$$

where G_{ST} is the etalon (standard) antenna gain and Γ_{ST} is the etalon antenna reflection coefficient.

When comparison method is used, it is especially important to know what parameter is indicated in etalon datasheet—G_r or G. The main advantage of this method is that there is no need to know:

- Probe gain;
- Transmission coefficient of entire cable system a_n;
- Reflection coefficients Γ_l, Γ'_p.

However, for the accurate gain determination, it is still desirable to know the reflection coefficients Γ_g, Γ_a, Γ_{ST}. The main disadvantage of this method is the necessity of the additional time-consuming etalon antenna scan. It is better to carry out measurements of the main scan for the AUT and the additional scan for the etalon antenna at the same day, because during this time the conditions under which measurements are taken (temperature, antenna position, and cables position) do not change significantly. Alteration of the measurement conditions may result in the gain determination error.Therefore, this method cannot be recommended if a large number of measurements should be carried out. It is advisable to choose the size of both scans based on the recommendations described earlier. As a rule, the scan steps in both coordinates are chosen the same for both scans. Scan area sizes can be different for AUT and standard scans. It is important to remember that if the etalon antenna has more spatially localized amplitude distribution than the AUT one, the scan areas sizes must be different. In practice, such a situation may occur if the antenna under investigation is a large-scale array and the reference antenna is a pyramidal horn.

Three-antenna method. To calculate the gain with three-antenna method, it is necessary to carry out a series of three measurements, in each of which, at the same power supplied to the input of the transmitting antenna, the power values at the output of the receiving antenna are measured. The measurements are carried out with the following antenna pairs: 1 (transmitting)–2 (receiving), 1 (transmitting)–3 (receiving), and 2 (transmitting)–3 (receiving). The gain of each antenna is obtained by solving a system of three equations with three unknowns:

$$\begin{aligned} G_{rA1} &= \sqrt{\frac{G_{rA1}G_{rA2}\cdot G_{rA1}G_{rA3}}{G_{rA2}G_{rA3}}}; \\ G_{rA2} &= \sqrt{\frac{G_{rA1}G_{rA2}\cdot G_{rA2}G_{rA3}}{G_{rA1}G_{rA3}}}; \\ G_{rA3} &= \sqrt{\frac{G_{rA2}G_{rA3}\cdot G_{rA1}G_{rA3}}{G_{rA2}G_{rA1}}}, \end{aligned} \tag{3.202}$$

where G_{rA1}, G_{rA2}, G_{rA3} are the realized gains of antennas.

Products under the radicals are determined by the results of the above-mentioned measurements:

$$G_{rA1}G_{rA2} = \frac{P_{A12}}{P_{TX}}\left(\frac{4\pi R}{\lambda}\right)^2;$$
$$G_{rA1}G_{rA3} = \frac{P_{A13}}{P_{TX}}\left(\frac{4\pi R}{\lambda}\right)^2; \quad (3.203)$$
$$G_{rA2}G_{rA3} = \frac{P_{A23}}{P_{TX}}\left(\frac{4\pi R}{\lambda}\right)^2,$$

where $P_{A12}, P_{A13}, P_{A23}$ are the powers received by the antennas in three measurements; P_{TX} is the power transmitted to the path; and R is the distance between antennas.

For example, the ratio of the powers in a spherical coordinate system can be expressed as follows:

$$\frac{P_A}{P_{TX}} = \frac{\left|E_\varphi(\theta_o, \varphi_o)\right|^2 + \left|E_\theta(\theta_o, \varphi_o)\right|^2}{a_n^2}. \quad (3.204)$$

In case of three-antenna method for the planar near-field measurements, it is more convenient to use expressions based on (3.199) and (3.200) instead of (3.203)

$$\mathbf{G_{rA1}G_{rA2}} = \left(\frac{4\pi}{\lambda^2}\right)^2\mathbf{M_r}\left|\sum_{\mathbf{x_m},\mathbf{y_m}} \frac{\mathbf{b'_{012}(x_m,y_m,d)}}{\mathbf{a_n}}\mathbf{e^{ik_{x0}x_m+ik_{y0}y_m}\Delta x\Delta y}\right|^2;$$
$$\mathbf{G_{rA1}G_{rA3}} = \left(\frac{4\pi}{\lambda^2}\right)^2\mathbf{M_r}\left|\sum_{\mathbf{x_m},\mathbf{y_m}} \frac{\mathbf{b'_{013}(x_m,y_m,d)}}{\mathbf{a_n}}\mathbf{e^{ik_{x0}x_m+ik_{y0}y_m}\Delta x\Delta y}\right|^2; \quad (3.205)$$
$$\mathbf{G_{rA2}G_{rA3}} = \left(\frac{4\pi}{\lambda^2}\right)^2\mathbf{M_r}\left|\sum_{\mathbf{x_m},\mathbf{y_m}} \frac{\mathbf{b'_{023}(x_m,y_m,d)}}{\mathbf{a_n}}\mathbf{e^{ik_{x0}x_m+ik_{y0}y_m}\Delta x\Delta y}\right|^2.$$

The main advantage of the three-antenna method is the possibility to measure the gain in the absence of information about the measured antenna characteristics. Since when measuring in the near field, one of the antennas is a probe with a wide RP, the implementation of this method at a planar near-field facility is not entirely justified. It is due to the fact that with limited transverse dimensions of the scan area, it is not possible to intercept all the energy emitted by the probe when using it as a transmitting antenna, which leads to errors in the gain determination. In addition, there is a likelihood of the Gibbs effect occurrence caused by a field discontinuity at the edge of the scanning surface, while the amplitude–phase distribution at scan area transforms to the far-field RP. In this case, the calculated antenna RP can have oscillations, which will be an additional source of errors in determining the gain. This is especially true for low-frequency antennas, where it is often difficult to achieve a field decrease at the boundaries of the scan area even up to −20 dB relative to the maximum of the amplitude distribution. Thus, it is most reasonable to use three-antenna method at spherical near-field facilities, since in this case it is possible to intercept all the energy of the antenna with any radiation pattern (it should be remembered that for spherical scanning, the measurement radius is the distance from the intersection point of the rotation axis of the antenna rotary support to the phase center of the probe; see Chap. 5). To illustrate the above, the results of gain measurement with three-antenna method on the spherical near-field facility are depicted in Figs. 3.14 and 3.15. Etalon horn 1 has datasheet gain error ±0.5 dB, and etalon horn 2 (R&S) has gain error ±1 dB.

As follows from the graphs shown in Figs. 3.14 and 3.15, measured gain of all three antennas coincides well enough with the characteristics declared by the manufacturers in datasheets. The main drawback of the three-antenna method is the big time consuming. It is optimal to use this method to find the characteristics of

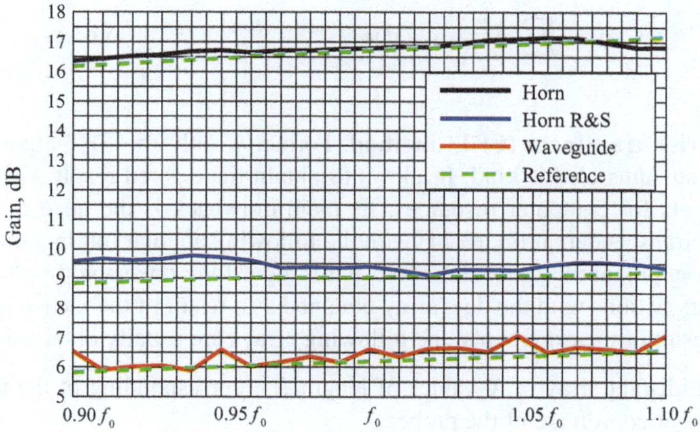

Fig. 3.14 Gain calculated with the three-antenna method at the spherical near-field facility

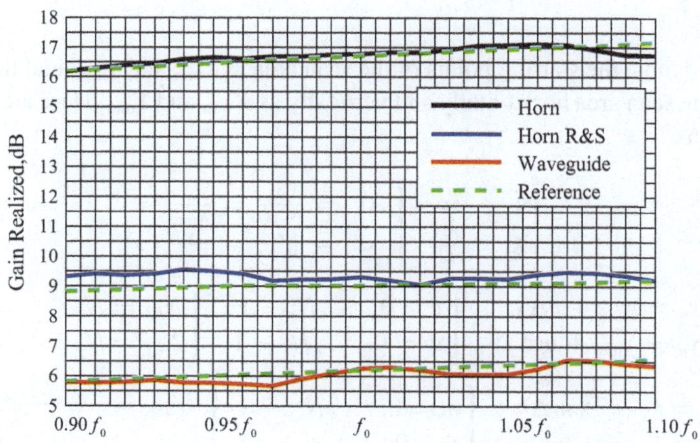

Fig. 3.15 Realized gain calculated with the three-antenna method at the spherical near-field facility

measuring probes in order to create a set of their calibration characteristics, which can then be used to find the gain of the AUTs using the direct method.

3.7 Methods for Numerical Calculation of Integral Sums

Let us discuss in detail the methods for finding sums of the form:

$$A^{(m,n)}(\theta,\varphi)=\sum_{x_v,y_w}E(x_v,y_w,d)\mathrm{e}^{ik_x(\theta,\varphi)x_v+ik_y(\theta,\varphi)y_w+ik_zd}\Delta x\Delta y \tag{3.206}$$

Fast Fourier transform (FFT) method. Formerly, only the FFT algorithm was used to find sums of this kind. It allows to obtain the desired result with minimal requirements for computer resources. Its main drawback is the need to solve the system of transcendental Eq. (3.74) with the following angular interpolation. Let us take a close look at this algorithm. To use this algorithm, one should perform some preliminary actions with the data array obtained as a result of the measurement.

As a result of measurements, the following arrays are usually obtained:

- An equidistant array of X values of length N_x, corresponding to the horizontal movement coordinate of the probe;
- An equidistant array of Y values of length N_y, corresponding to the vertical movement coordinate of the probe;
- An array of dimension $\left(N_x,N_y\right)$ of the measured amplitudes and phases at each scan point.

Let us denote the starting points of the scan area as $x_{\min}$ and $y_{\min}$, and the dimensions of the scan area horizontally and vertically as $X_{\min}$ and $Y_{\min}$. Here are the main expressions:

$$\begin{cases}\tilde{k}_x=\frac{k_x}{k}\\ \tilde{k}_y=\frac{k_y}{k}\end{cases}\quad\begin{cases}\Delta\tilde{k}_x=\frac{\lambda}{N_x\Delta x}=\frac{\lambda}{X_{\max}}\\ \Delta\tilde{k}_y=\frac{\lambda}{N_y\Delta y}=\frac{\lambda}{Y_{\max}}\end{cases} \tag{3.207}$$

$$\begin{cases}x_v=x_{\min}+v\Delta x;\\ y_w=y_{\min}+w\Delta y;\end{cases}\quad\begin{cases}v=0,\ldots,N_x-1\\ w=0,\ldots,N_y-1\end{cases}\quad\begin{cases}N_{kx}=N_x\\ N_{ky}=N_y\end{cases} \tag{3.208}$$

$$\begin{cases}\tilde{k}_{xm}=\tilde{k}_{x\min}+m\Delta\tilde{k}_x;\\ \tilde{k}_{ym}=\tilde{k}_{y\min}+n\Delta\tilde{k}_y;\end{cases}\quad\begin{cases}m=0,\ldots,N_{kx}-1\\ n=0,\ldots,N_{ky}-1\end{cases}\quad\begin{cases}\tilde{k}_{x\min}=\frac{-(N_{kx}-1)}{2}\Delta\tilde{k}_x;\\ \tilde{k}_{y\min}=\frac{-(N_{ky}-1)}{2}\Delta\tilde{k}_y;\end{cases} \tag{3.209}$$

$$\begin{cases}k_{xm}x_v=2\pi\left(\frac{x_{\min}\tilde{k}_{x\min}}{\lambda}+\frac{\Delta x\tilde{k}_{x\min}}{\lambda}+\frac{x_{\min}m}{N_x\Delta x}+\frac{vm}{N_x}\right);\\ k_{yn}y_w=2\pi\left(\frac{y_{\min}\tilde{k}_{y\min}}{\lambda}+\frac{\Delta y\tilde{k}_{y\min}}{\lambda}+\frac{y_{\min}n}{N_y\Delta y}+\frac{wn}{N_y}\right).\end{cases} \tag{3.210}$$

The solution to the expression (3.206) using the FFT algorithm has the form:

$$A_x^{(m,n)} = \frac{\Delta x \Delta y}{4\pi^2} \exp(i\psi) \cdot \exp(i\psi_s) \cdot \mathrm{IFFT}\{E_x^{(m,n)}\}, \tag{3.211}$$

where

$$\psi = 2\pi\left(\frac{x_{\min}\tilde{k}_{x\min}}{\lambda} + \frac{y_{\min}\tilde{k}_{y\min}}{\lambda} + \frac{x_{\min}m}{N_x\Delta x} + \frac{y_{\min}n}{N_y\Delta y}\right), \tag{3.212}$$

$$\psi_s = 2\pi\frac{d}{\lambda}\sqrt{1 - \left(\tilde{k}_{x\min} + \frac{m\lambda}{N_x\Delta x}\right)^2 - \left(\tilde{k}_{y\min} + \frac{n\lambda}{N_y\Delta y}\right)^2}. \tag{3.213}$$

The FFT algorithm itself, by definition, is expressed as:

$$\mathrm{IFFT}\{E_x^{(m,n)}\} = \sum_{v=0}^{N_x-1}\sum_{w=0}^{N_y-1} \tilde{E}_x^{(v,w)} \exp\left\{2\pi i\left(\frac{vm}{N_x} + \frac{wn}{N_y}\right)\right\}, \tag{3.214}$$

where

$$\widetilde{E}_x^{(v,w)} = E_x^{(v,w)} \exp\left\{2\pi i\left(\frac{\Delta x\tilde{k}_{x\min}}{\lambda}v + \frac{\Delta y\tilde{k}_{y\min}}{\lambda}w\right)\right\}. \tag{3.215}$$

After obtaining the result by using (3.211), it is necessary to solve the system of transcendental Eq. (3.174) and interpolate for the desired grid of angles θ and φ.

Matrix method. This method can be implemented in programming languages designed for highly efficient matrix calculations, for example, MATLAB® or Octave. Nowadays, the problem of computer memory lack has disappeared. In this connection, it has become possible to solve (3.206) using direct matrix multiplication. This allows to eliminate errors associated with the solution of transcendental equations and subsequent interpolation. In this case, vectors of required θ values with size $[1, L_\theta]$ and vector of required φ values with size $[1, L_\varphi]$ directly substitute to (3.174). Thus, matrices $\hat{k}_x$, $\hat{k}_y$ and $\hat{k}_z$ with size $[L_\theta, L_\varphi]$ are obtained. Then, $\hat{k}_x$ and $\hat{k}_y$ matrices are transformed (reshaped) into vectors $\vec{k}_x$ and $\vec{k}_y$ of size $[1, L_\theta * L_\varphi]$. The vector of X coordinates with size $[L_x, 1]$ multiplies with the vector $\vec{k}_x$ of size $[1, L_\theta * L_\varphi]$ that results in matrix $X\vec{k}_x$ of size $[L_x, L_\theta * L_\varphi]$. Each element of matrix $X\vec{k}_x$ is multiplied by an imaginary unit, after which it serves as a power for corresponding element of a matrix of the same size composed of e (the base of the natural logarithm) numbers (elementwise exponentiation). The result is the $e^{iX\vec{k}_x}$ matrix of size $[L_x, L_\theta * L_\varphi]$.

Similarly, the multiplication of the vector of Y coordinates with size $[L_y, 1]$ with the vector $\vec{k}_y$ of size $[1, L_\theta * L_\varphi]$ results in the matrix $Y\vec{k}_y$ of size $[L_y, L_\theta * L_\varphi]$. The next step is the matrix $\mathrm{e}^{iY\vec{k}_y}$ of size $[L_y, L_\theta * L_\varphi]$ calculation. After that, the matrix of the complex field values E_{xy} (obtained as a result of measurement) with size $[L_y, L_x]$ needs to be multiplied with the matrix $\mathrm{e}^{iX\vec{k}_x}$ of size $[L_x, L_\theta * L_\varphi]$. The result is the matrix $\hat{E}_{xy}\mathrm{e}^{iX\vec{k}_x}$ of size $[L_y, L_\theta * L_\varphi]$. Thus, as a result of these operations, summation over one of the coordinates is performed. Then, the matrix $\hat{E}_{xy}\mathrm{e}^{iX\vec{k}_x}$ should be elementwise multiplied by matrix $\mathrm{e}^{iY\vec{k}_y}$ that results in the matrix of size $[L_y, L_\theta * L_\varphi]$. After that, this matrix should be summarized over the L_y dimension, thus obtaining the vector $\vec{A}_{mn}$ of size $[1, L_\theta * L_\varphi]$. This vector should be reshaped to the matrix $\hat{A}_{mn}$ of size $[L_\theta, L_\varphi]$. And finally, elementwise multiplication of matrix $\hat{A}_{mn}$ by the matrix $\mathrm{e}^{ik_z d}$ of size $[L_\theta, L_\varphi]$ and multiplication by two scalars $\Delta x, \Delta y$ lead to the result. The entire procedure can be written as follows:

$$\left(\mathrm{reshape}\left(\mathrm{sum}\left[(\mathrm{e}^{iY*\vec{k}_y}.*(\hat{E}_{xy}*\mathrm{e}^{iX*\vec{k}_x}))\right]\right).*\mathrm{e}^{ik_z d}\right)*\Delta x*\Delta y \quad (3.216)$$

where (.*) is the elementwise matrix multiplication and (*) is the matrix multiplication.

3.8 Measurement Errors When Scanning on a Planar Surface

This section only lists the main measurement errors of planar near-field facilities and indicates possible ways to reduce them. Methods for calculating these errors are not considered in this section. Some calculation formulas can be found in [10, 13].

Errors in determining the radio characteristics of antennas at the planar near-field facility can be divided into two groups:

1. Methodical errors;
2. Measuring instruments errors.

The first group (methodical errors) includes the following errors:

1.1 Due to a limited scan plane, on which amplitudes and phases of the near field of the AUT are measured;
1.2 Due to the sampling the amplitude and phase values of the AUT near field;
1.3 Due to the finite size of the probe;
1.4 Due to the reading of the measured parameters without stopping the probe during its movement.

The group (1.1) errors are due to the fact that the CTC measurement is not performed on an infinite planar surface. This error becomes negligible if the boundaries of the measurement area are determined by the level of field decay relative to the maximum by 40 dB at the selected distance from the aperture of the AUT to the probe.

The group (1.2) errors depend on the step value of the discrete sequence of points where the values of amplitudes and phases of CTC are measured. The discrete step of measurement for values of amplitude and phase of CTC should not exceed 0.5 λ. If this discrete step value does not exceed 0.1 λ, then the errors become negligible. The practical discrete step value for coordinates of measurement points depends on the measuring equipment parameters and the capabilities of the computer used.

The group (1.3) errors are due to the fact that the real probe measures the values of amplitudes and phases of CTC not in points, but integrates them on a certain surface. To eliminate this error, it is necessary to use an elementary oscillator as the measuring probe. If a directional antenna with known RP is used as a probe, its influence on the measurement results can be taken into account analytically.

The group (1.4) errors are due to the need to interpolate the measurement results since with the probe continuously moving along the scan plane, the measurement of $|E_{\tau x}|$, $\varphi_{\tau x}$ values occurs at discrete points when the coordinate controller sends a trigger signal to the receiver. Since the sweep time of the receiver is not infinitesimal, and the velocity of the probe is not absolutely stable, the measurements at each frequency are slightly shifted in space. Therefore, the collected experimental data must be subjected to preliminary processing consisting in the alignment of the discrete steps of the coordinate grid by means of a linear (or some other) approximation of the parameters $|E_{\tau x}|$ and $\varphi_{\tau x}$ dependence on spatial coordinates. The arrays "corrected" in this way are the initial data for further calculations.

Instrumental errors include:

2.1 Ampliphasemeter error;
2.2 Errors due to microwave generator parameter instability;
2.3 Errors introduced by the paths of the reference and measuring channels;
2.4 Errors caused by the probe feeder deformation in the process of moving the probe across the scan surface;
2.5 Errors caused by multiple reflections of the radio signal between the antenna under test and the probe;
2.6 Errors caused by reflection from surrounding objects;
2.7 Errors of the equipment for recording coordinates of the point on the scan surface.

When vector network analyzers are used in near-field facilities, errors of the groups (2.1), (2.2), (2.3) are determined by the parameters of the instrument and specified in its datasheet.

The group (2.4) errors are determined by the parameters of the high-frequency cables used. The approximate assessment of these errors can be carried out using datasheets for these cables. A more accurate assessment can only be done experimentally.

The group (2.5) errors are inevitable, but can be reduced by the following methods:

- Using a small-sized probe;
- Using radio absorbing materials;
- Measurement of amplitude–phase distribution on two similar surfaces with the subsequent compensation by calculation of the error caused by multiple reflections.

The group (2.6) errors can be reduced by performing the following operations:

- Using radio absorbing materials to "mask" the surrounding objects during the measurements;
- Taking measurements in anechoic chambers;
- Taking measurements in the time domain.

The group (2.7) errors are determined by the parameters of the equipment used and should be indicated in the datasheets of the corresponding devices.

References

1. Korn GA, Korn ThM (1961) Mathematical handbook for scientists and engineers. McGraw Hill Book Company, New York
2. Paris DT, Leach WM, Joy EB (1978) Basis theory of probe-compensated near-field measurements. IEEE Trans Antennas Propag AP-26(N3):373–379
3. Johnson W (1988) An examination of the theory and practices of planar near-field measurement. IEEE Trans Antennas Propag 36(6):746–752
4. Slater D (1991) Near-field antenna measurement. Artech House, Inc., 310 p
5. Yaghjian (1986) An overview of near-field antenna measurements. IEEE Trans Antennas Propag AP-34(N1):30–45
6. Joy P (1972) Spatial sampling and filtering in near-field measurements. IEEE Trans Antennas Propag AP-20(N.3):253–261
7. Kerns DM (1976) Plane-wave scattering-matrix theory of antennas and antenna-antenna interactions: formulation and applications. J Res Natl Bureau Stand B Math Sci 80B(1):40–49
8. Joy EB, Marshall Leach W, Rodrigue GP, Paris DT (1978) Applications of probe-compensated near-field measurements. IEEE Trans Antennas Propag 26(3):379–389
9. Gregson S, McCormick J, Parini C (2007) Principles of planar near-field antenna measurements. In: IET electromagnetic waves series, vol 53. The Institution of Engineering and Technology, London
10. Visser HJ (1989) Theory of planar near-field measurement. FYSISCH EN ELEKTRONISCH LAB TNO THE HAGUE (NETHERLANDS), № fel-89-b273, Dec 1989
11. Stuzman WL, Thiele GA (1998) Antenna theory and design, 2nd edn. Wiley, New York, 79 p

12. Ango A (1964) Mathematics for electrical and radio engineers. Nauka, Moscow, 772 p (in Russian)
13. Newell AC, Ward RD, Mcfarlane EJ (1988) Gain and power parameter measurement using planar near-field techniques. IEEE Trans Antennas Propag 36(6):792–803
14. Sushkevich VI (1967) Irregular linear waveguide systems. Soviet Radio, Moscow, 295 p (in Russian)

Chapter 4
Planar Near-Field Facility: Electrical and Mechanical Parts, Software

4.1 Generalities

There are several schemes of planar near-field facilities. The most widespread is the classical one, in which the tower moves along horizontal slide-rails. Tower has horizontal guide-rails along which, in turn, probe carriage moves. Device under test is installed vertically, in a plane parallel to the plane of movement of the tower. In addition to the classical construction scheme, in practice there is a bridge scheme. In this case, the DUT is located horizontally and aperture upward, and the probe mounted on a Π-shaped structure passes over it [1]. The choice of a particular scheme is primarily determined by the intended measurement objects and the dimensions of the room where the facility will be located. The bridge scheme is suitable for measuring large-sized objects of complex configuration that are difficult to orient properly. For example, antennas mounted on spacecraft. Various types of planar facilities are shown in Fig. 4.1.

4.2 Facility Block Scheme

Planar near-field facility block scheme is shown in Fig. 4.2.

Let us consider the purpose and composition of the block scheme elements.

Probe horizontal movement mechanism is intended to ensure the movement of the tower, with the probe fixed, along the horizontal slide-rails. The tower is mounted on a movable trolley, on which there are rollers that allow it to roll along horizontal slide-rails. For the movement of the tower, three main types of drives are used:

- Friction drive (the movement of the trolley is provided by the transmission of torque from electric motors to the rollers of the trolley and the adhesion of these rollers to the slide-rails due to the frictional force that increases under the influence of the weight of the tower);

V. S. Kalashnikov et al., *Near-Field Antenna Measurements*, Springer Aerospace Technology, https://doi.org/10.1007/978-981-33-6436-3_4

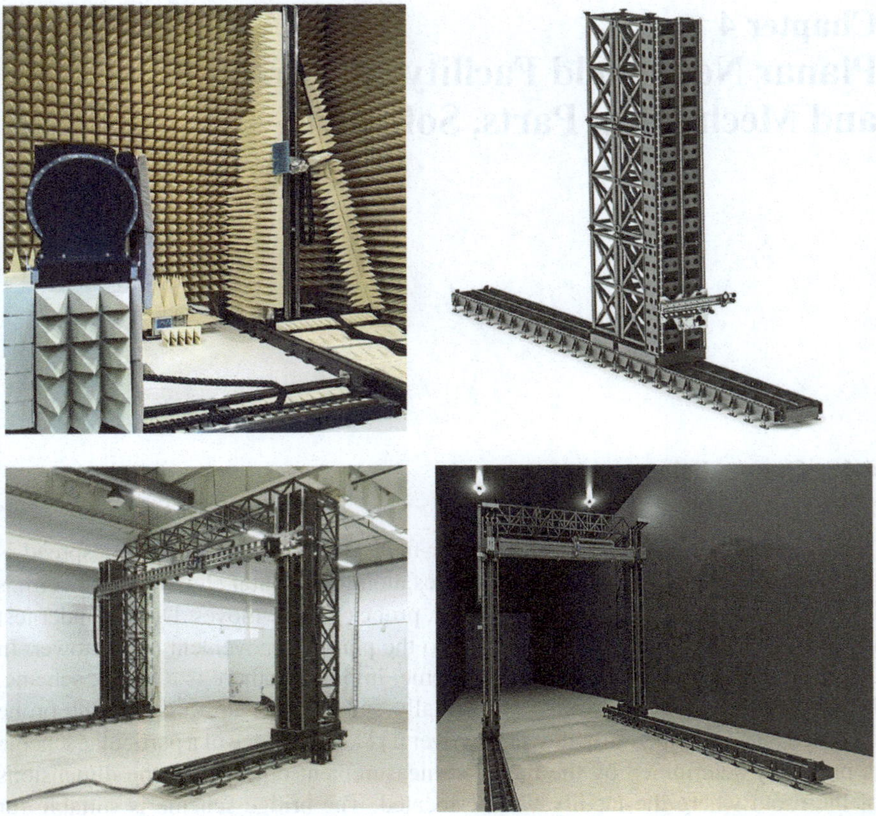

Fig. 4.1 Planar near-field facilities of various types

- Gear drive (the movement of the trolley is provided by a gear rack and gear pair, in which the rack is stationary and fixed on a facility base, and the gear is mounted on the trolley and rotates by an electric motor);
- Chain drive (the movement of the trolley is provided by winding of chain attached to the trolley onto a rotating barrel).

One of the important requirements for horizontal movement is smoothness. The fulfillment of this requirement provides a reduction in dynamic loads on the tower and, therefore, on the probe, which is fixed on it. In addition, smooth running provides a reduction in the measurement error of the phase of the complex transmission coefficient that occurs when the probe oscillates. To fulfill this requirement, the motion path is divided into three parts: the acceleration zone, the uniform motion zone, and the deceleration zone. The disadvantage of the chain drive is its low smoothness in the uniform motion zone. A drawback of the friction drive is the slippage of the rollers when dirt gets on the rails or when the rollers jump due to the unevenness

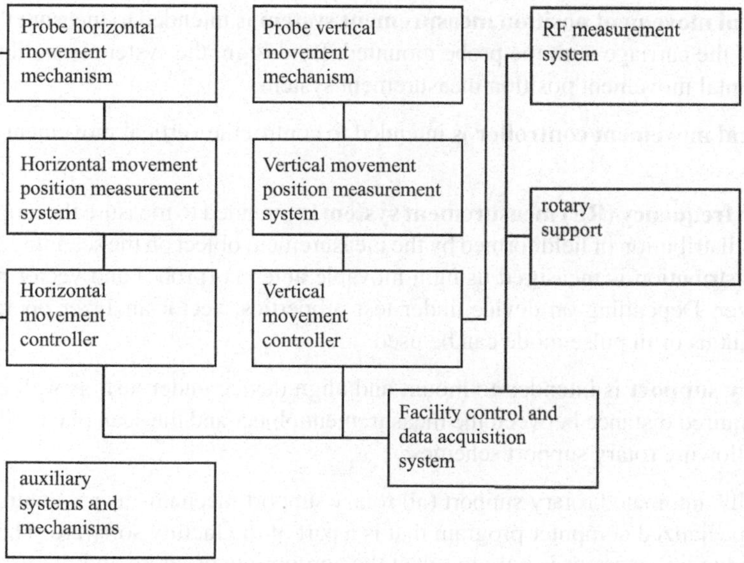

Fig. 4.2 Facility block scheme

of the rails. Moreover, if information about the position of the tower is taken from linear encoders mounted on the roller, slippage will result in a positioning error.

Horizontal movement position measurement systems differ in the type of encoder used: absolute and incremental one. A specificity of the incremental system is its dependence on power source. When the power is turned off, the current coordinate is reset. Subsequently, after turning on the power, the current position of the scanner becomes initial, i.e., zero, and the coordinates will be counted from this point. In order to be able to compare the results of measurements of the same object taken at different times, the scanner must return to the starting point after each measurement.

Horizontal movement controller is intended to control the horizontal movement stepper motor. The controller is based on a microprocessor and, as a rule, is produced by the manufacturer together with the motor.

Probe vertical movement mechanism is intended to move the carriage, with the probe mounted, along the vertical guide-rails. For the movement of the carriage, two main types of drives are used:

- Movement along a gear rack mounted between vertical guide-rails;
- Chain drive.

For smooth carriage motion, the vertical movement mechanism is equipped with a counterweight.

Vertical movement position measurement system is intended to measure the position of the carriage with the probe mounted. By design, the system is similar to the horizontal movement position measurement system.

Vertical movement controller is intended to control the vertical movement stepper motor.

Radio frequency (RF) measurement system is intended to measure the amplitude–phase distribution of field formed by the measurement object on the scanning surface. The distribution is measured using a movable antenna (probe) and vector network analyzer. Depending on device under test properties, vector analyzer operating in continuous or in pulse mode can be used.

Rotary support is intended to mount and align device under test, as well as to set the required distance between the measurement object and the scan plane. There are the following rotary support schemes:

- Fully automated rotary support (all rotary support mechanisms are controlled by a specialized computer program that is a part of the facility software. The role of the facility operator is only to select the appropriate program and run it);
- Semi-automated rotary support (some of rotary support mechanisms are controlled by a specialized computer program that is a part of the facility software, and the rest are controlled by the facility operator in manual mode);
- Manually controlled rotary support (all rotary support mechanisms are controlled by operator in manual mode).

Motion control and data acquisition system are intended to:

- Form the facility movement parameters.
- Send commands to the controllers of the facility motors.
- Receive information from motor controllers through feedback circuits.
- Form the RF measurement system parameters.
- Collect output data from the RF measurement system.
- Receive information from linear and angular encoders.
- Preprocess the output data of the RF measurement system.
- Store files with processed data for LAN share, printing or copying to a flash drive.

Auxiliary systems and mechanisms do not directly participate in the measurement process, but significantly facilitate the working conditions of the operator. These may include:

- System of axial probe movement, intended to set the required distance between the probe and the AUT;
- System of probe rotation providing the required position of the plane of polarization (for a linearly polarized probe) without remounting it;
- System for remote distance measurement (used to align large-sized measurement objects);
- Auxiliary system for tower position correction;

- Speakerphone system (for large-scale chambers, or if control room is located outside the chamber);
- Television surveillance system (for visual control of the facility and the measurement object);
- Television system on the probe bar (for visual control of the probe positioning).

4.3 Mechanical Part of the Facility

The mechanical part of the facility includes horizontal guide-rails, tower, motors, and vertical guide-rails. To reduce the influence of external destabilizing factors (vibrations from technological equipment, machine tools, passing vehicles), the facility basement should be independent of the building basement. The requirements for roughness of the outer surface of the foundation are specified by the equipment manufacturer and can be in the range from 5 to 25 mm with normal shrinkage. Caped cable channels for facility control and power cables and device under test control and power cables should be provided in the floor. Horizontal and vertical guide-rails can have various designs depending on the manufacturer. In most cases, horizontal guide-rails are polished rectangular rails mounted on the frame, and less often they are cylindrical ones. Carriages are installed on the guide-rails. They have ready-made flanges with mounting holes, a nipple for supplying lubricant to the rolling elements (ball or roller bearings), and brushes that prevent debris from entering the bearing assembly.

As a rule, the frame is installed on the foundation with the help of special anchor bolts, which are also used for its adjustment. Figure 4.3 shows the installation diagram of these bolts. The frame can be all-in-one-piece or consists of separate sections,

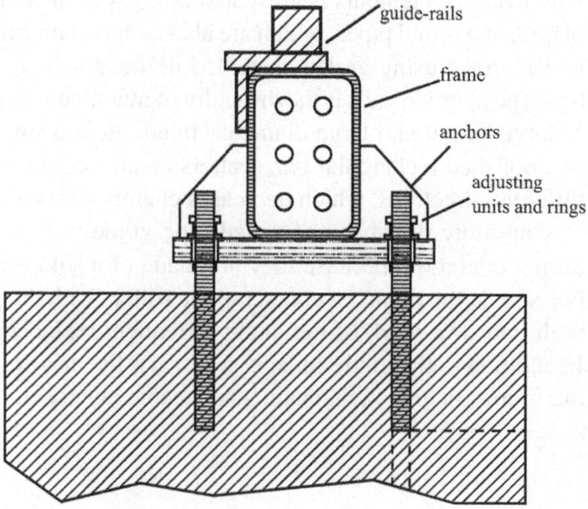

Fig. 4.3 Variant of the horizontal guide-rails and the frame mounting to basement

Fig. 4.4 Welded steel frame with installed elements

which can be made in one of three options: molten cast iron, less often molten steel; steel welded, more rarely built-up; and aluminum built-up.

The aluminum frame is assembled from drawn aluminum profiles, which are fastened to each other with threaded joints. Using an aluminum profile allows one to get a lightweight and at the same time quite rigid structure, easy to assemble, and reliable in operation. The use of molten frames provides high structural rigidity, but increases its weight (compared with other options). Steel welded structures occupy an intermediate position between the above options. Welded steel frame with installed elements is presented in Fig. 4.4. In some cases, guide-rails are made of polished round pipes, which are also mounted on frame. Horizon plane adjustment is performed using gaskets installed in the attachment points. The disadvantage of this type of guide-rails is the difficulty of manufacturing if one wants to produce pipes of large length and large diameter. In addition, a smooth polished pipe, in contrast to a polished rectangular bar, scatters plane electromagnetic waves incident on it in different directions, which decreases chamber anechoic characteristics.

One more drawback of round pipe guide-rails is the rapid wear of the rollers during operation, because they are made of a softer material than the pipe, and the pressure at the points of contact of the rollers with the pipes is very high. As a result, with intensive work of the facility, frequent replacement of the rollers is required. In addition to the guide-rails, if necessary, the drive mechanisms are also located on the frame—a gear rack, chain gears, cable layers, etc.

Probe tower is intended to install the vertical guide-rails of the probe carriage, the carriage itself, and the drive of this carriage. The tower should satisfy two contradictory requirements: minimum weight and maximum rigidity. To meet these requirements, the tower, as a rule, is made of aluminum profile. The general view of the tower, vertical guide-rails, and attachment points are shown in Figs. 4.5 and 4.6.

Direct current stepper motors and synchronous motors are usually used in near-field facilities. Stepper motors, as compared to synchronous ones, have lower speed and dynamic characteristics and lower power, but they also have lower price. Facilities that do not experience high dynamic loads can be equipped with asynchronous motors with feedback. The most perfect technical solution for facilities today is a "direct" drive. Its principle of action is the direct conversion of electromagnetic energy into

Fig. 4.5 Tower and carriage general view

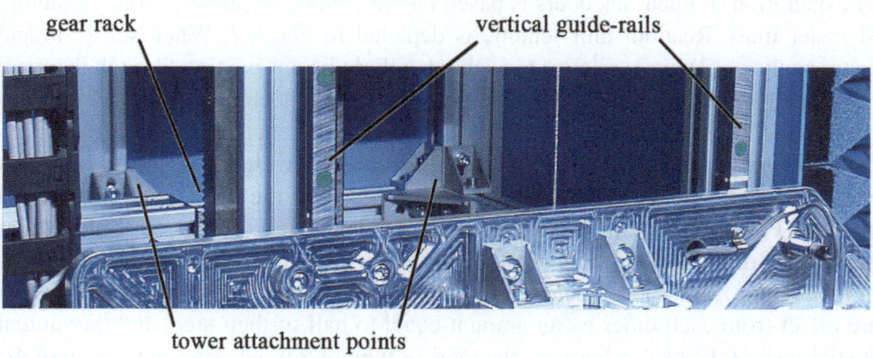

Fig. 4.6 Vertical guide-rails, mounting points

mechanical energy of linear motion. Such a drive provides the best performance in almost all parameters: accuracy, acceleration and deceleration dynamic, speed, and repeatability. In linear drives, there are no rotating parts subject to wear and friction; therefore, over time, the characteristics of the drive practically do not change. The linear drive has only one drawback—the high price.

4.4 Linear Encoders

Linear encoders (sensors) are intended to measure object position and send it to controller or indicator. There are two types of sensors: absolute and incremental ones. Absolute linear displacement sensors for each value of the object position associate the value of the numerical equivalent, which is formed at the output of the sensor (usually in the form of a digital code signal). This one-to-one correspondence is preserved both during continuous motion of the tower and during its stepped motion. Thus, the code value is not lost after turning the sensor off and on. This is a big advantage of absolute linear encoders compared with incremental ones. There are two types of linear encoders: optoelectronic and magnetic ones.

Let us consider principle of operation of optoelectronic sensors based on "SKBIS Ltd" materials [2]. A specificity of linear optoelectronic raster encoders is the use of a linear scale, which is a carrier of regular and coding rasters, as a distance measure. The possibility of putting raster lines with sub-micrometer accuracy to materials with a specified linear expansion coefficient, as well as the stability of their geometric position, led to the creation of encoders of 3–4th accuracy classes.

4.4.1 Principle of Operation of the Optoelectronic Linear Encoders

The operation of linear encoders is based on the method of optoelectronic scanning of raster lines. Readout unit scheme is depicted in Fig. 4.7. When scale (1) and raster analyzer (3) move relative to each other, light beams passing through them are modulated and then received by corresponding photodetectors. The raster scale has two parallel tracks: the regular raster track and the reference mark one.

The raster analyzer has four windows of incremental readout B, $\bar{B}$, A, $\bar{A}$ and reference mark window C. Figure 4.8 shows the structure of the window of incremental readout. Positions of four aforementioned windows are coordinated with regular raster track of the scale. The steps of the rasters in the windows are equal to the steps of the regular raster of the scale (20 or 40 μm). In each pair of windows, the rasters are offset from each other by an amount equal to half of their step, and the mutual spatial shift of the rasters between the pairs of windows is a quarter of the step of the rasters.

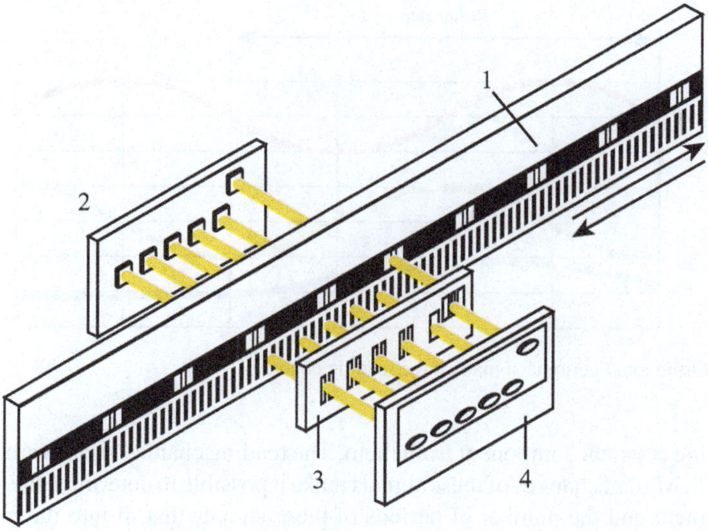

Fig. 4.7 Readout unit scheme. 1—Readout unit scale, 2—photodetector board (silicon photodiodes), 3—raster analyzer, 4—LED board (infrared emitters)

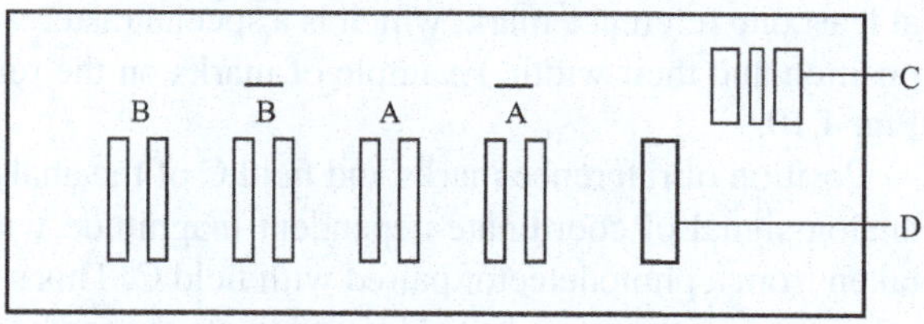

Fig. 4.8
B, $\bar{B}$, A, $\bar{A}$—windows of incremental readout,
C—reference mark window,
D—transparent window

Next to raster windows, transparent window D is located. The position of reference mark C is coordinated with the reference mark track of the scale.

Readout unit (read head) of the linear encoder solves two tasks: the first—maps the optical raster and the code signal, the second—readout, processing, and analysis of the optical information. The carriage rigidly connected to the analyzer solves the first task. The carriage is in constant contact with the scale (through rolling bearings), which makes relative movement of the scale and the analyzer possible. The second task is solved by the photodetector board (2), LED board (4) installed on the same carriage, and the board for extracting and processing movement information located in the read head case. The LED board contains six emitting diodes, which provide illumination of the corresponding analyzer windows, and the receiving areas of six photodiodes of the board spatially matched with them (2). The reading channel designed in this way allows forming of two orthogonal periodic signalsI_A andI_B,

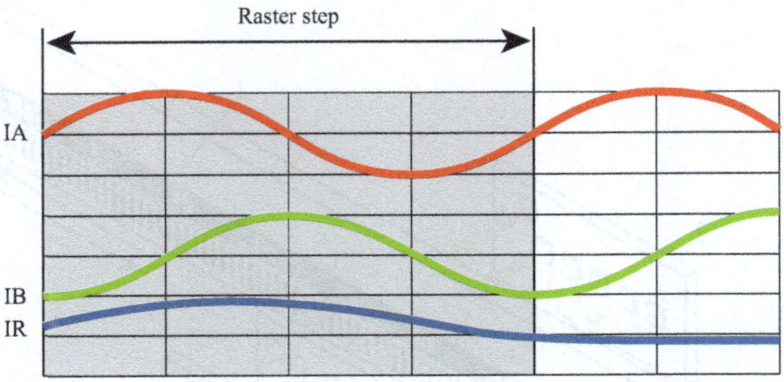

Fig. 4.9 Orthogonal periodic signals of the reading channel

eliminating constant component from them. The reading channel signals are depicted in Fig. 4.9. Mutual changes of these signals make it possible to determine the direction of movement and the number of periods of these signals that fit into the raster step directly proportional to the magnitude of this movement. Special methods of signal I_A and I_B processing make it possible to control movement with a resolution that is much smaller than the regular raster period.

To enable to set user reference point, a reference mark track is used, which contains at least one reference mark, which is a special raster with a given function for stroke position and their width. Example of marks on the readout unit scale is depicted in Fig. 4.10.

Position of reference marks and field C of the analyzer is shown in Fig. 4.11. An analog signal of coordinate-dependent magnitude with a pronounced maximum is taken from a photodetector paired with field C. This signal is used by the movement information processing board to coordinate the reading unit to the reference point of the encoder. In addition, the value of the reference signal specified by the photodetector corresponding to the analyzer window D is taken into account in this process. In LIR-7 … LIR-10 linear encoders, the moving element during operation can be either a scale or a read head. The user makes the choice during encoder mounting.

The logical development of linear encoders using a sequence of coordinate-coded reference marks is the creation of quasi-absolute linear encoders in which the track

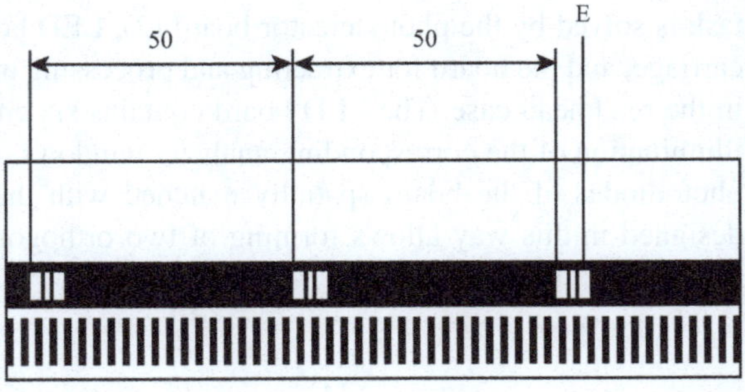

Fig. 4.10 Markers on readout unit scale

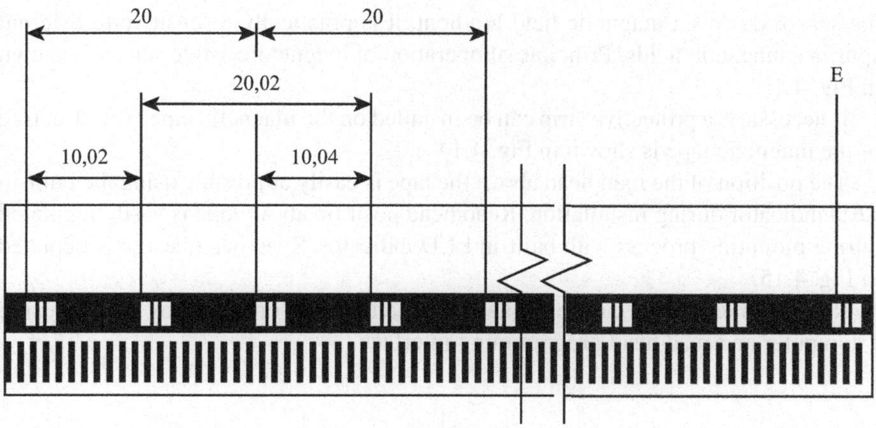

Fig. 4.11 Coordinate-coded location of reference marks *E*

of reference marks is replaced by a code track. This makes it possible to determine the coordinate of the position of the readout unit relative to the scale after turning on the power and a relative shift of 0.5 mm, which provides quasi-absolute linear sensor with the property of an absolute position sensor throughout the entire operation cycle until it is turned off. Let us consider the principle of operation of magnetic sensors using the LM10 sensor manufactured by RLS (Slovenia) as an example. One of the read head designs is depicted in Fig. 4.12 [3].

A differential magnetoresistive sensor detects a magnetic field gradient above a magnetic tape and converts it into analogous sine and cosine signals. These analogous signals are then converted inside the read head using an interpolator, which provides a resolution up to 1 μm. To ensure the reliability of measurements, it is necessary that the gap between the sensor and the tape does not exceed ¾ of the distance from the north to south pole of the magnetic tape. Within such a gap, the amplitude of the sine and cosine signals is stable. The tolerance does not exceed 10%. Since

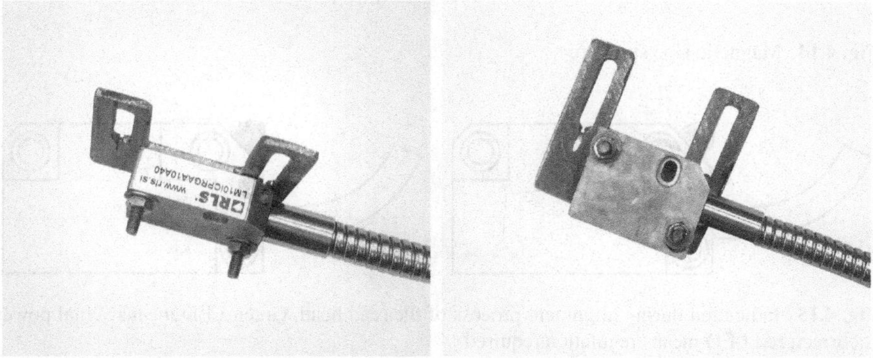

Fig. 4.12 LM10 read head

the sensor detects a magnetic field gradient, it is practically insensitive to external spurious magnetic fields. Principle of operation of magnetoresistive sensor is shown in Fig. 4.13.

If necessary, a protective strip can be installed on the magnetic tape. The structure of the magnetic tape is shown in Fig. 4.14

The position of the read head above the tape is easily adjustable using the built-in LED indicator during installation. Read head position above tape is easily regulated during mounting process with built-in LED indicator. Read head design is depicted in Fig. 4.15.

Read head resolution can be regulated on manufacture or chosen by the user using a computer and read head programming interface.

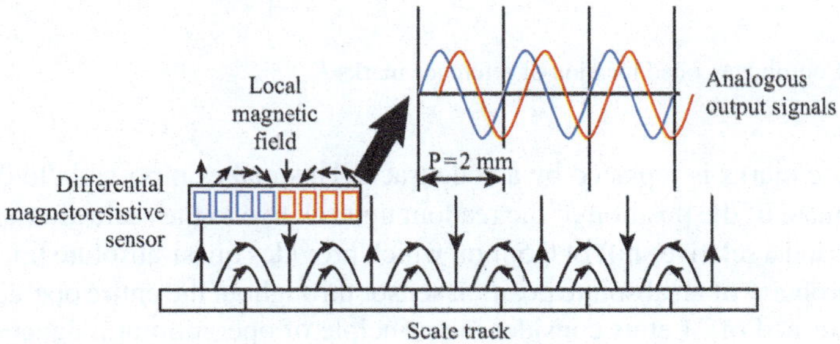

Fig. 4.13 Principle of operation of magnetoresistive sensor

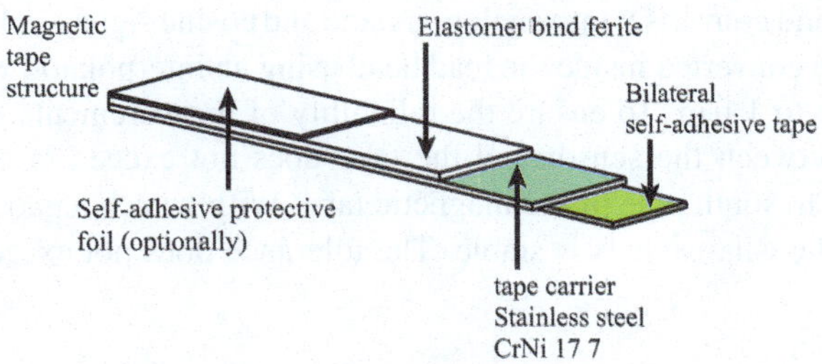

Fig. 4.14 Magnetic tape structure

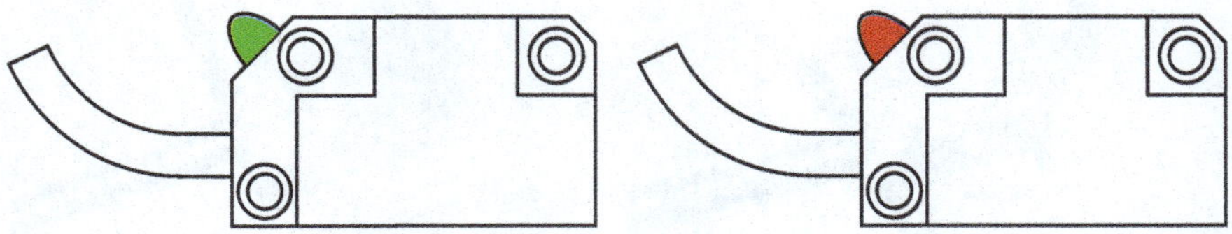

Fig. 4.15 Indication during alignment process of the read head. Green LED means signal power is correct; red LED means regulation required

4.4.2 *Output Codes of the Absolute Encoders*

1. Parallel output with gray code (resolution up to 12 digits). The code structure is depicted in Fig. 4.16.
   All bits (D0-D11) of the angle code of the encoder spindle are simultaneously present on the sensor output bus. Table 4.1 shows the characteristics of the signals.
2. Parallel binary code with fixed signal (resolution up to 12 digits). The code structure is depicted in Fig. 4.17.
   LE is the input control signal. On the negative edge of the LE signal, the encoder spindle position code is fixed. As long as the LE level is zero, bits D0 to D11 are present on the output bus.Z is the high impedance state,n is the number of bits, andt_5 andt_6 are transition times, depending on the type of output signal.
3. Parallel byte code (resolution up to 14 digits). Code structure is depicted in Fig. 4.18.
   The code of the spindle position is fixed at the moment of the negative OE1 signal edge. At zero level OE1 on the output bus, there are bits D0-D7 of the spindle position code, and at OE2 zero level—the remaining bits. OE1 and OE2 are input control signals. Z is the high impedance state. t_1, t_2, and t_3 are 300 ns. Table 4.1 shows the signal parameters.
4. Parallel byte code (resolution 15 and 16 digits). The code structure is depicted in Fig. 4.19.

Fig. 4.16 Parallel gray code

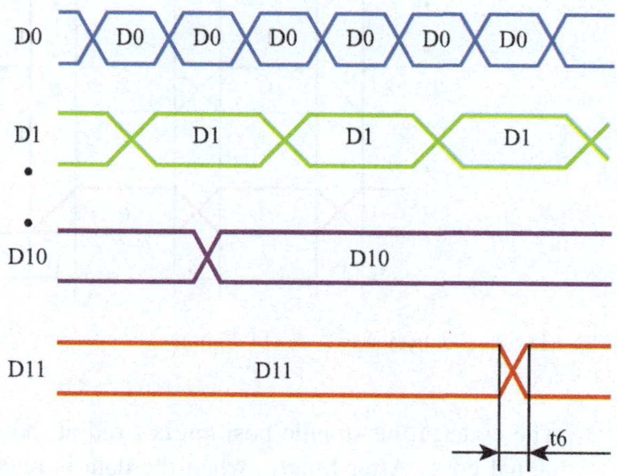

Table 4.1 Time of transition from Z-state and back

Output signal type	Standard TTL (TP) (μs)	Open collector (OC) (μs)	Open emitter (OE) (μs)
t_5	≤0.3	≤0.3	≤0.5
t_6	≤0.3	≤0.5	≤2.0

Fig. 4.17 Parallel gray code with fixed signal

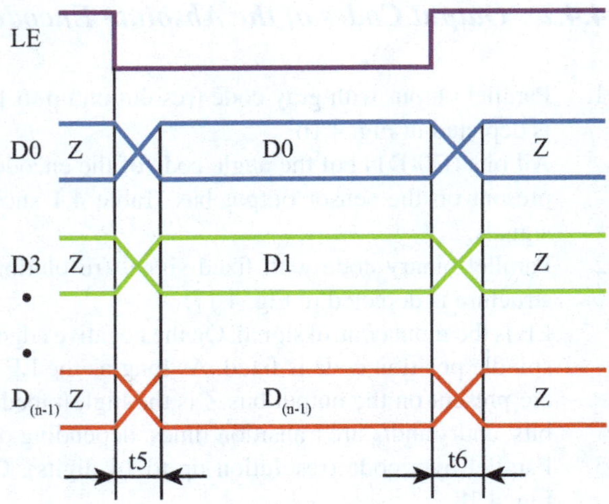

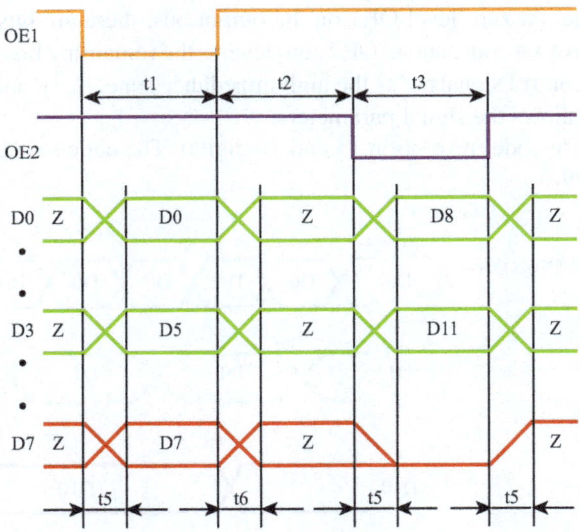

Fig. 4.18 Parallel byte gray code (14 digits)

The code of the spindle position is fixed at the moment of the negative OE1 signal edge. After timet_1 (when the data is ready to be sent to the bus), the output signal GD is set to zero. At a zero level of the data availability signal GD and a zero level OE1, the bits D0-D7 of the spindle position code are present on the output bus, and at zero levels GD and OE2 there are D8-D15. OE1 and OE2 are input control signals. GD is the output signal.Z is the high impedance state.

5. Serial SSI (regular, without additional ALARM bit). The code structure is presented in Fig. 4.20.

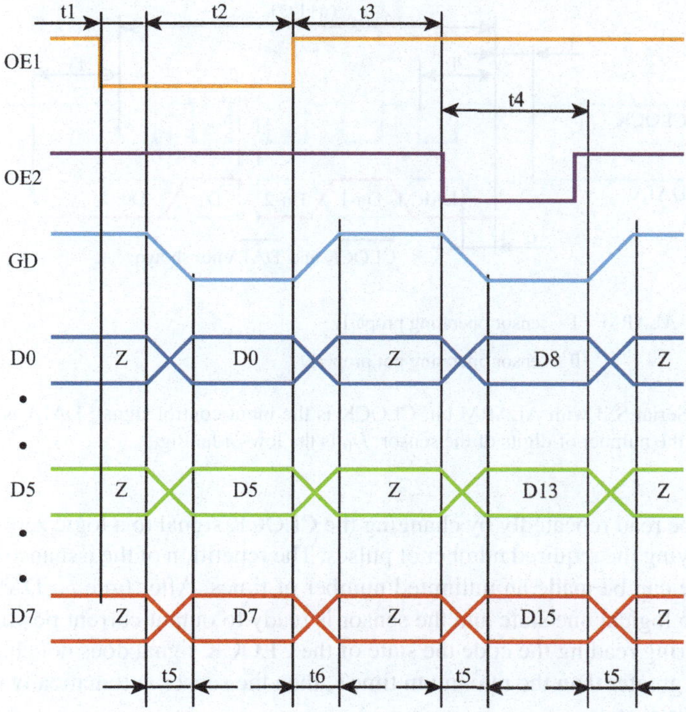

Fig. 4.19 Parallel byte gray code

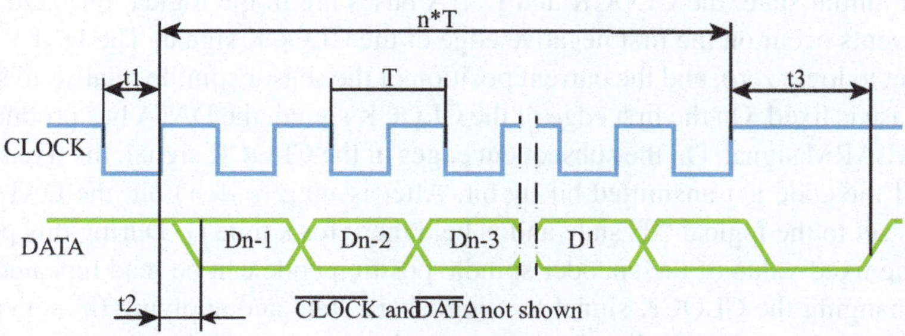

Fig. 4.20 Serial regular SSI. CLOCK—input control signal; DATA—output signal; *n* is the number of digits of the sensor

In initial state, the CLOCK and DATA buses are in the logical one state. On the first negative edge of the CLOCK signal, the value of the sensor spindle position code is fixed in the sensor buffer. On the subsequent positive edges of the CLOCK signal, the fixed value of the code is transmitted bit by bit, starting with the high-order one. After issuing *n* bits, the DATA line is set to the logical "0" state and is held there for a time t_3. During this period, a fixed code value

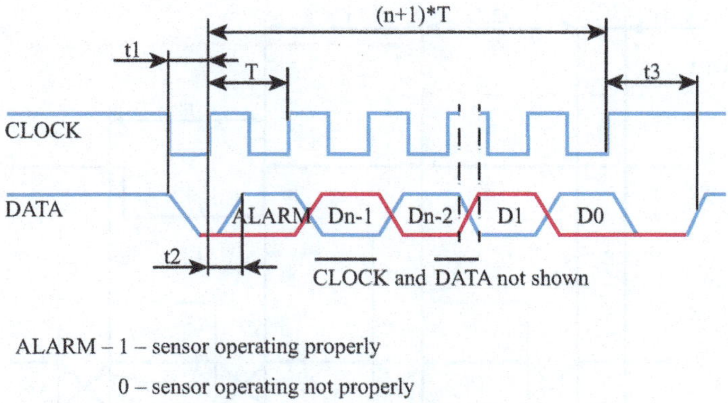

Fig. 4.21 Serial SSI with ALARM bit. CLOCK is the input control signal; DATA is the output signal; n is the number of digits of the sensor; D_0 is the low-order digit

can be read repeatedly by changing the CLOCK signal to a logic zero state and applying the required number of pulses. The repetition of the issuance of a fixed value can be made an unlimited number of times. After time t_3, DATA line is set to logical one state and the sensor is ready to output current position value. If during reading the code the state of the CLOCK signal does not change for a time greater than the maximum time$T$, then the sensor automatically returns to its initial state.

6. Serial SSI a (with additional ALARM bit). The code structure is presented in Fig. 4.21.
   In initial state, the CLOCK and DATA buses are in the logical one state. Two events occur on the first negative edge of the CLOCK signal: The DATA bus is set to logic zero, and the current position of the sensor spindle relative to sensor case is fixed. On the first edge of the CLOCK signal, the DATA bus produces an ALARM signal. On the subsequent edges of the CLOCK signal, the fixed value of the code is transmitted bit by bit. After issuing $(n + 1)$ bit, the DATA line is set to the logical "0" state and is held there for a time t_3. During this period, the fixed value of the encoder spindle position code can be read repeatedly by changing the CLOCK signal to a logic zero state and applying $(n + 1)$ pulse. After time t_3, DATA line is set to logical one state and the sensor is ready to output current position value. If during reading the code the state of the CLOCK signal does not change for a time longer than t_3, then the sensor automatically returns to its initial state.

4.5 Rotary Supports

The rotary support is intended for DUT installation and its aligning relative to the measurement plane of a planar near-field facility. The design choice of the rotary support is determined by the parameters of the object being measured at the facility. These parameters include the mass of the DUT, its overall dimensions, and the presence of a moving base in the structure of the DUT. To perform the alignment of the DUT, the rotary support should be able to move in at least three planes: in azimuth, roll, and elevation. Additionally, to ensure the necessary conditions for choosing the size of the scan area, the support rotary can be able to change the height of the DUT and to ensure the required distance to the scan plane can be installed on the rails, providing movement in the plane perpendicular to the scan plane.

According to the degree of automation, rotary supports are divided into three groups: non-automated; semi-automated; and automated.

Non-automated rotary supports are manually controlled by the facility operator or technician. Semi-automated rotary supports are partially controlled manually and partially using positioners. Automated rotary supports are controlled with positioners. Steering can be carried out from a working place of the operator or from the remote control. An example of the simplest non-automated rotary support for lightweight antenna is the tripod shown in Fig. 4.22.

The tripod rotary support allows to manually rotate DUT in azimuth. In addition, the adjustment of the DUT by elevation and roll can be carried out by changing the length of the tripod supports.

The product manufactured by "Smitek" company depicted in Fig. 4.23 can serve as an example of fully automated rotary support that provides controlled rotations of the DUT in azimuth, elevation, and roll angles as well as axial moving [1].

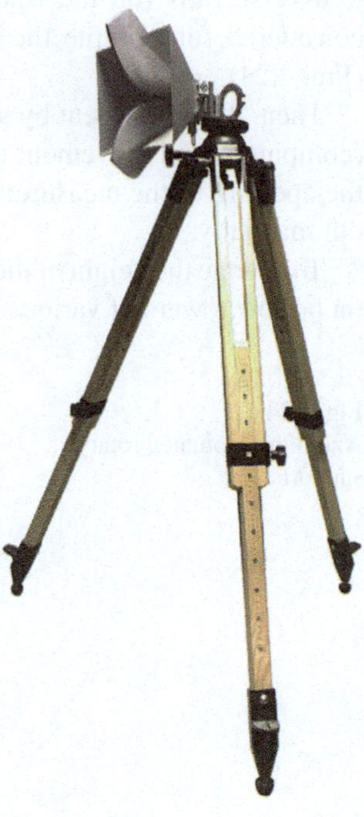

Fig. 4.22 Tripod with antenna installed

Fig. 4.23 Fully automated rotary support

Semi-automated rotary supports are used for lightweight antennas. As an example of a semi-automated rotary support, any automated rotary support mounted on the transverse rails (on the frame of Fig. 4.4) on a non-automated platform can be considered, for example, the azimuth-automated rotary support of "Smitek" company (Fig. 4.24).

Then, the adjustment by azimuth is carried out by commands from the controller (computer), and movement to set the required distance between the scan plane and the aperture of the measurement object, adjustment by roll, and elevation is carried out manually.

To change the height of the measurement object, the rotary support can be installed on hollow towers of various heights having flanges for attaching the rotary support.

Fig. 4.24 Azimuth-automated rotary support

A set of towers may be shipped with the rotary support. The guide-rails for moving the rotary support are identical to the guide-rails of the probe tower.

4.6 Probes and Variants of Its Mounting

When near-field measurement is carried out, as a rule, the AUT emits radiation, and the auxiliary antenna, called the probe, receives the signal at a number of points on the scanning surface (plane). If the facility does not include non-reciprocal microwave devices (e.g., an amplifier or an isolator), then the principle of reciprocity is valid for the antenna and the probe; i.e., during measurements, the AUT can operate both in reception and transmission modes, and the probe, respectively, and in transmission and reception modes. When measuring the amplitude and phase of the tangential component of the electric field strength vector excited by the AUT on the scanning surface, the probe introduces certain distortions into the received signal. One of the reasons for these distortions is the difference between the probe RP and the RP of an ideal isotropic radiator. This reason can be eliminated if information about the probe RP is introduced into the algorithm of preliminary processing of the measurement results. Such an information can be determined theoretically or by special measurements. The requirements for probes are often contradictory and cannot be fully satisfied. These requirements depend on the properties of the measuring equipment. Nevertheless, several general rules can be formulated for all types of probes:

- The probe should keep its characteristics during long time period.
- The probe should keep its characteristics in various positions (taking into account gravity).
- The probe should have a good reference to the coordinate system.
- The probe should have low cross-polarization level (it is desirable that the cross-polarization level be no more than −30 dB).
- The probe should introduce minimal distortion into the measured field (as one of the consequences of this requirement, probe should match feeder line good).
- The probe should provide a low level of the reflected (scattered) signal in the angular region in front of the antenna under study.
- Probe attachment elements, including its feeder lines, must be well covered by a radio absorbing material or designed in such a way as to guarantee minimal scattering toward the AUT.

There is also a list of optional requirements:

- The probe should be wideband.
- The probe should be dual polarized.
- The probe should have standard RF connector.
- The probe should be lightweight.

Probe with wide RP is desirable for planar facility. The requirements for the probe RP are depicted in Fig. 4.25.

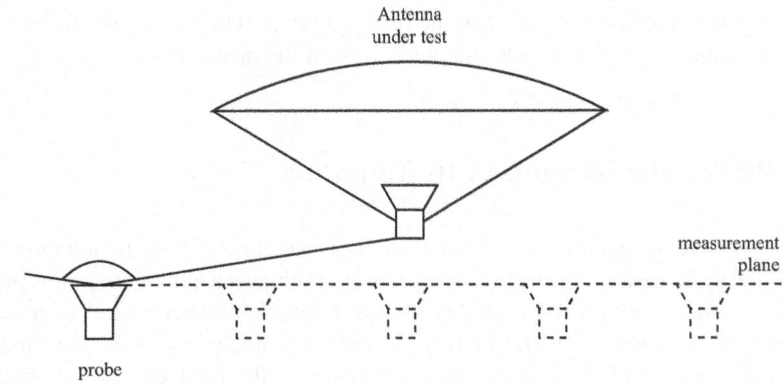

Fig. 4.25 Requirements for the probe RP

4.6.1 Dipoles and Loop Probes

Small (compared to the wavelength) dipoles and small loop antennas are often used as probes for near-field measurements. Their main advantage is that they introduce minimal distortion into the measured field and that their radiation patterns are very wide, almost like that of an isotropic radiator, which makes it possible to skip probe correction when processing the measurement results. The disadvantages of small dipoles include a low input impedance, which makes it difficult to match these antennas with a standard 50-Ω feeder line.

4.6.2 Open-Ended Waveguide Probes

The open-ended waveguide (rectangular, square, or circular one) is often used as a probe for frequencies above 600 MHz. The open-ended standard rectangular waveguide excited by the fundamental mode TE_{10} has the following advantages as a probe: It is relatively wideband (approximately 1:1.5), simple to manufacture, and inexpensive. Drawbacks are operating at single linear polarization and sufficiently high level of backward radiation. The latter drawback makes it necessary to carefully design the absorbing coating for the probe attachment elements, since a slight change in the placement of the radio absorbing material due to dismantling and reinstallation may lead to significant changes in the probe RP. There are a number of open-ended waveguide probe manufacturers on the market. Waveguide probes and their RPs are presented in Figs. 4.26, 4.27, 4.28, and 4.29.

To save weight of the rectangular waveguide probe, slots can be cut in the middle of the wide side of the waveguide (Fig. 4.29).

Comparing the rectangular waveguide probe, the circular one is more narrowband; however, it has two major advantages, making it the best probe for planar near-field

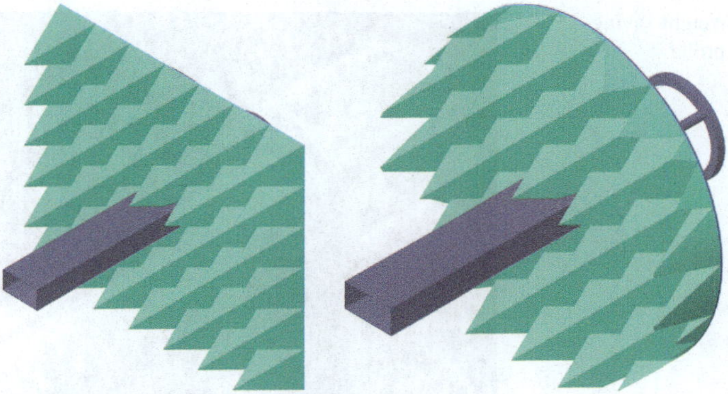

Fig. 4.26 Open-ended waveguide probes with a rectangular and circular absorber *on the screen* for backward scattering reducing

Fig. 4.27 Radiation pattern of the open-ended rectangular waveguide probe

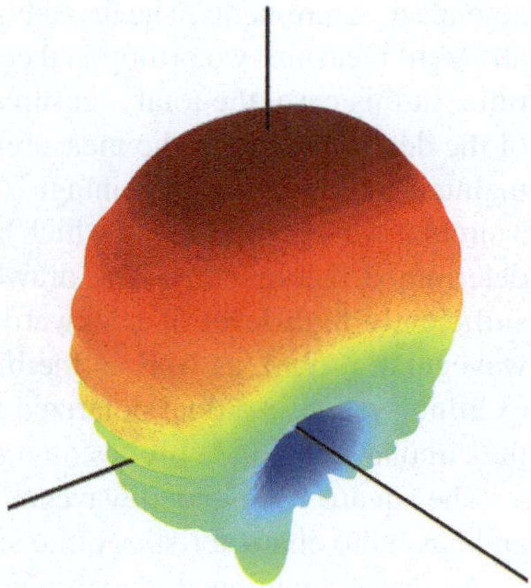

Fig. 4.28 Open-ended rectangular waveguide probe with absorber glued

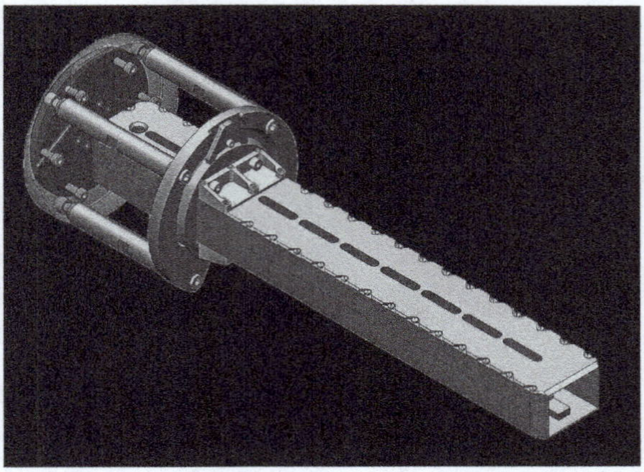

Fig. 4.29 Weight-saving waveguide probe

antenna measurements. The first advantage is it can be designed dual polarized, which allows to measure two orthogonal components of the electric field vector at the same time. In this case, the total measurement time is reduced and the complete identity of the determination of the measurement point coordinates for both polarizations is maintained. The second advantage is this probe is excited by the fundamental mode of a circular waveguide (TE_{11}), which has symmetrical structure, and probe RP is easily determined theoretically. The drawbacks of the circular waveguide probe include sufficiently high level of backward radiation (the same as that of the rectangular waveguide probe), as well as the high complexity of the design of the waveguide exciting system in a dual polarized variant. There are a number of manufactures of the circular waveguide probes on the market.

The square open-ended waveguide probe can also be designed as dual polarized and has radio characteristics quite similar to that of the circular waveguide probe.

As it was mentioned above, one of the drawbacks is waveguide probes are the big backward radiation. This radiation can be reduced by adding one or few choke grooves around the aperture (Fig. 4.30).

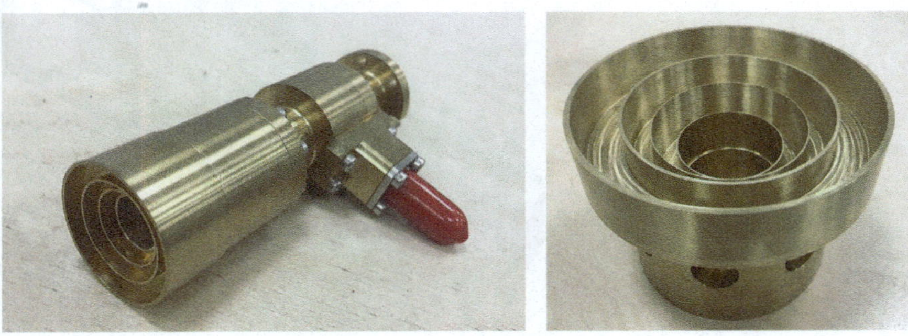

Fig. 4.30 Quarter-wavelength choke groove probes (Smitek company)

The use of choke grooves, however, increases the probe backscattering in direction to AUT. In addition, a single choke groove is a narrowband structure. The use of several choke grooves with shifted resonant frequencies allows to expand the band of the choke structure, but causes an increase in backscattering toward the AUT. The widening of the frequency band of waveguide probes can be achieved by manufacturing them on the basis of broadband waveguides: of Π-shaped, H-shaped, or quadruple-ridge ones. However, in comparison with regular waveguide probes, they are much more complicated to manufacture and have a high cost, and their radio characteristics can only be determined experimentally or by numerical simulation.

In Figs. 4.31 and 4.32, the designs of wideband waveguides are shown. Using

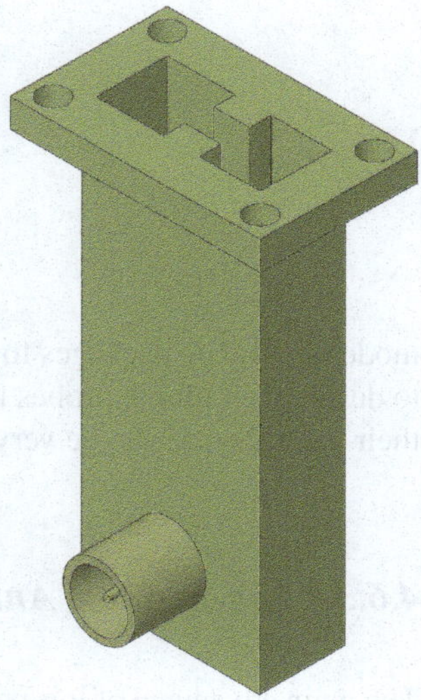

Fig. 4.31 H-shaped (double-ridge) waveguides

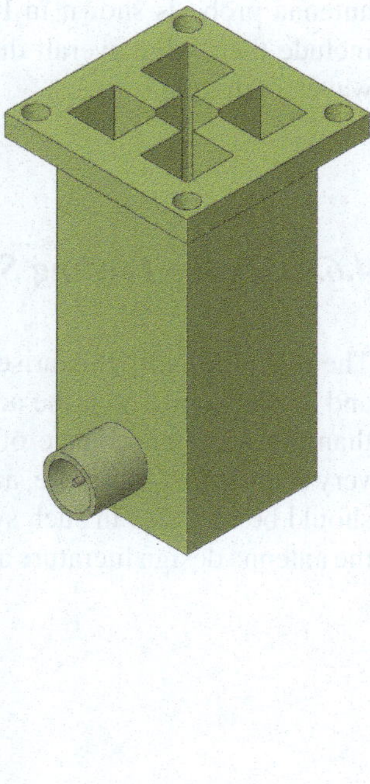

Fig. 4.32 Quadruple-ridge rectangular waveguide

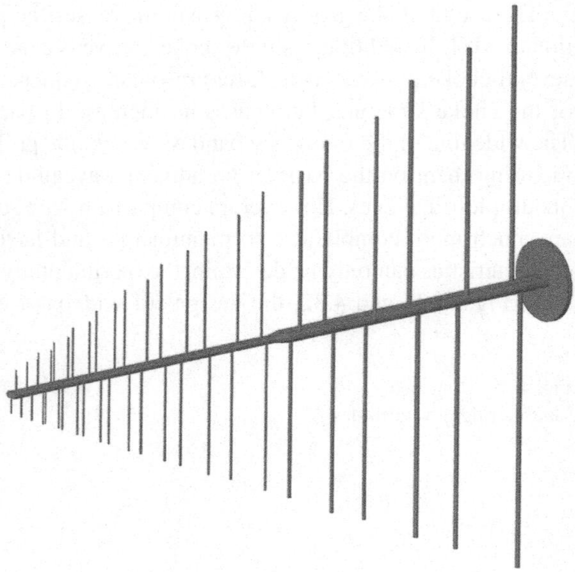

Fig. 4.33 Log-periodic antenna probe

modern software packages for mathematical modeling of microwave devices allows to design the optimal probes for given operating conditions; however, the process of their manufacture can be very complicated, and the price can be very high.

4.6.3 Log-Periodic Antenna Probes

The main advantage of log-periodic antennas is their ultra-broadband up to 1:25 and the ability to easily design and fabricate a dual polarized structure. Log-periodic antenna probe is shown in Fig. 4.33. The disadvantages of log-periodic antennas include their large overall dimensions and high mass, especially at the decimetric wavelength range.

4.6.4 Probe Feeding Systems

The main difficulty that arises when developing a feeding system for small dipoles and small loops is that the active component of their input impedance is much less than the wave impedance of standard feeder lines, and the reactive component is very significant. Therefore, active impedance transformers and non-dissipative stubs should be included in such systems. Specific engineering solutions can be found in the antenna design literature and are not discussed in this book. The feeding system of standard rectangular waveguide probe is usually designed as a waveguide to coaxial adapters. The feeding system of the circular waveguide can be implemented in a number of ways:

- Smooth or step transition from the circular waveguide to the rectangular one, which, in turn, is excited by the aforementioned waveguide to coaxial adapter;
- Asymmetric vibrator whose length is less than a quarter of the wavelength for the shortest wave from the probe range, located perpendicular to the inner wall of the circular waveguide;
- Two oppositely located asymmetric vibrators excited in antiphase (using an antiphase power divider, or with an in-phase power divider, but with cables, the electric length of which differs by 180° at the central frequency of the operating frequency range).

4.7 Ampliphasemeters

An ampliphasemeter is a device that can measure amplitude and phase of the complex transmission coefficient of a DUT connected between transmitting (generator) port and receiving port. At present, near-field facilities, as a rule, use vector network analyzers as a radio frequency measuring system (ampliphasemeter).

The appearance of vector network analyzer is presented in Fig. 4.34.

Two essentially different principles of vector network analyzer design are used: homodyne and heterodyne. Homodyne analyzers have only one generator. This generator produces an excitation signal, and the same signal is used as a reference signal for processing the response of the measured device. Most analyzers based on

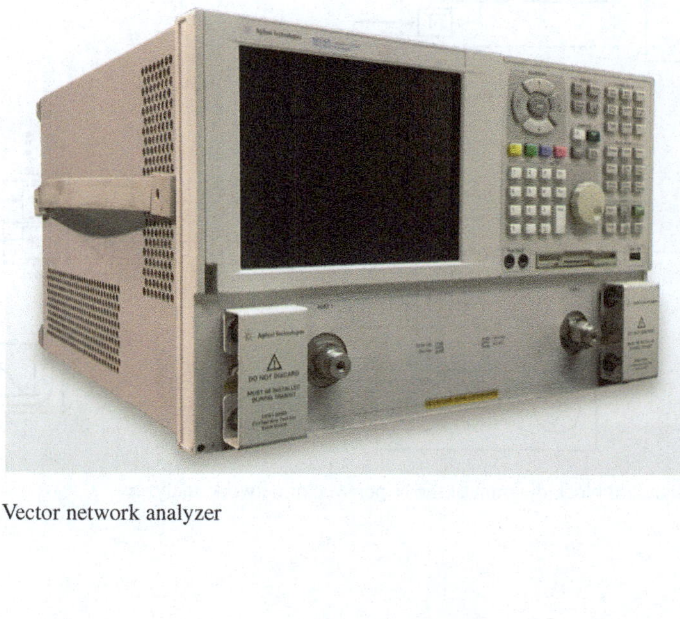

Fig. 4.34 Vector network analyzer

this principle are relatively inexpensive. However, due to various technical limitations inherent to them, analyzers constructed according to the homodyne principle are convenient only for simple applications and are not used for antenna measurement at near-field facilities. Heterodyne network analyzers have much broader capabilities and are used in modern near-field facilities. Let us consider the block diagram of the vector network analyzer. Block diagram of the N-port vector network analyzer is shown in Fig. 4.35. It consists of four main parts.

Test unit separates the incident and reflected waves at the test port. The separated waves arrive to the measuring channel and to the reference one. Electronic attenuators are used to measure the power level at the test port. To reduce the level of the power supplied to the test port, any step attenuators available in the generators can also be used.

Generator creates source RF signal, called excitation signal. An additional switch at the generator output directs excitation signal to one of the measuring ports, which in this case is considered as an active test port.

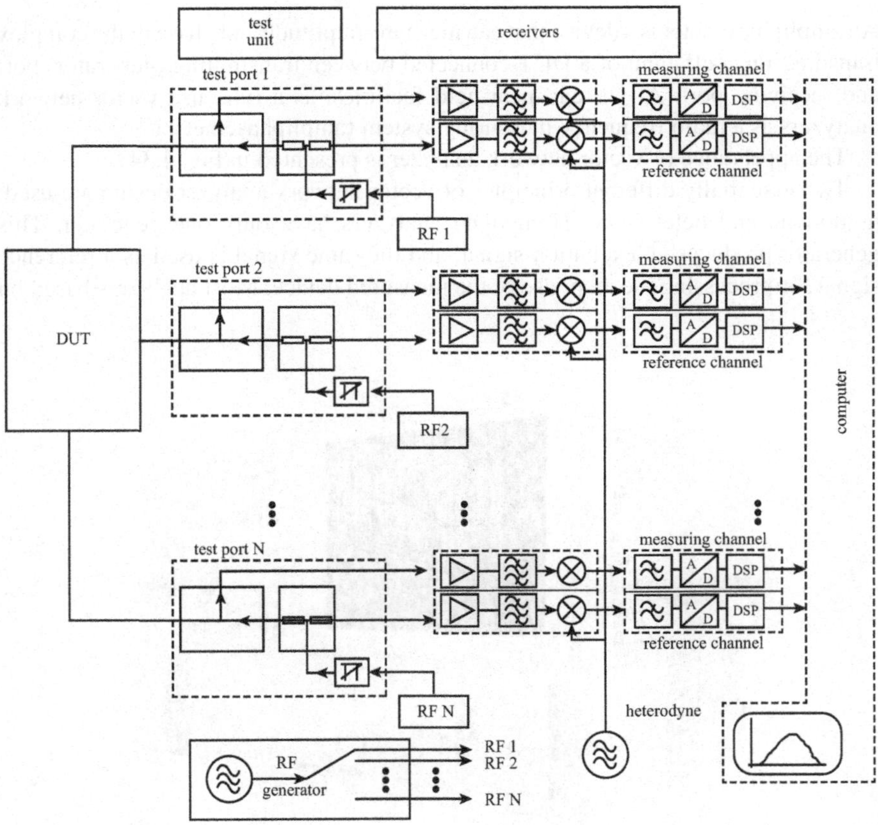

Fig. 4.35 Standard block diagram of the N-port vector network analyzer

Two separate receivers in each test unit are for the measuring channel and for the reference one. They are called measuring receiver and reference receiver. The receivers contain blocks shifting the RF signal to an intermediate frequency (implementation of the heterodyne principle). Next circuits of digital signal processing are located. At the outputs of the processing circuits, complex numbers corresponding to initial ("raw") measurement data are formed.

Computer is used for systematic error correction and measured data visualization. It also provides user interface and interfaces for remote control. The computer has special software, which is also called "firmware."

The vector analyzer can be remotely controlled in two ways—using the GPIB/Ethernet interface and using the COM/DCOM interface. Let us briefly consider the differences between these two methods.

Both GPIB and Ethernet interfaces implement the same command set based on the SCPI-1999 standard (Standard Commands for Programmable Instruments). This is a set of commands focused on the exchange of character messages. SCPI was developed by the SCPI Consortium (www.scpiconsortium.org). It is based on another standard IEEE488.2, except that SCPI can be used on any physical interface (GPIB, Ethernet, USB, RS-232), while IEEE488.2 is applicable only to GPIB. The main details of the SCPI standard are described below. More information on the SCPI standard can be found on the SCPI Consortium Web site. The main disadvantage of using SCPI technology is the need for a preliminary launch of the vector analyzer program; otherwise, the user program will not be able to connect to it. A less significant disadvantage is the limited functionality of the commands; i.e., not all vector analyzer parameters can be accessed from the user program. COM technology allows to start analyzer software remotely, provided that the vector analyzer is powered on and the operating system is running on it. COM technology is used when the user program and the analyzer software are installed and executed on the same computer (Fig. 4.36a). Distributed COM (DCOM) technology is used when the user program is installed and executed on a separate computer connected to the instrument computer via LAN.

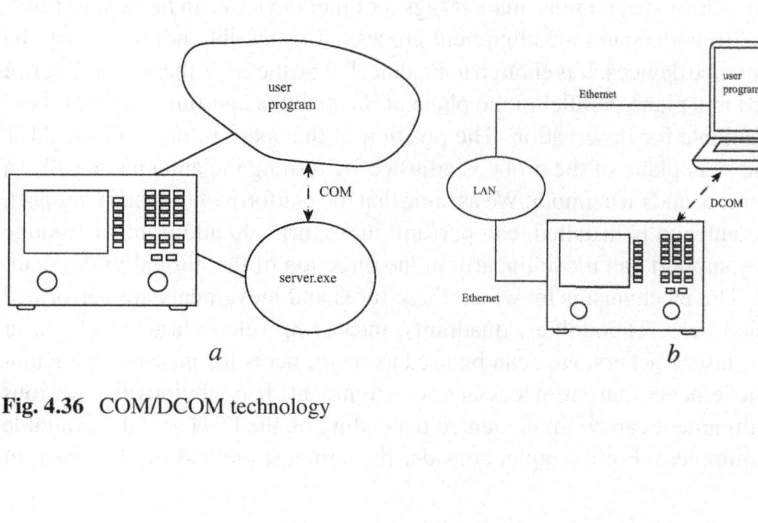

Fig. 4.36 COM/DCOM technology

The techniques and methods for writing user programs are the same for both technologies; the difference is that DCOM technology requires additional LAN configuration by the local network administrator. COM is a short for the *Component Object Model.* COM was designed in order to have two main features:

- The model provides a specification for interaction between binary modules written with various programming languages.
- The model determines the way of interaction between client application running on one computer with application server running on the same or another computer.

The following programming languages can be used to write user programs:

- Languages with built-in COM support, such as Visual Basic®, Delphi, Java;
- General-purpose programming languages, such as C and C++;
- Microsoft Office applications—Excel and Word because they contain the built-in programming language Visual Basic for Applications®;
- Program generators such as National Instruments LabVIEW®, Agilent VEE®, and MathWorks MATLAB®.

4.8 DUT Aligning

Alignment of the AUT (DUT) is carried out before the beginning of measurements when it is installed on the rotary support of the planar near-field facility. Alignment is carried out so that the plane of the aperture of this antenna installed and fixed on the rotary support is parallel to the scan plane (the probe movement plane). Alignment is particularly important in determining the radiation characteristics of narrowband antennas and antenna arrays, as it affects the accuracy of determining the angular position of the main and first side lobes of the RP of these objects. For alignment, it is necessary that the antenna design has been provided with special reference devices that allow one to determine the position of the plane of the radiating aperture of the antenna. The ends of special pins, marker tags, or other devices can be used as reference devices. To understand the alignment process, it is possible not to specify the design of reference devices. It is enough to assume that on the AUT there are reference points located in a plane parallel to the plane of the antenna aperture, and that these points are available for observation. The position of the aperture plane of the AUT relative to the scan plane of the probe is adjusted by turning the antenna on roll, on angle of elevation, and on azimuth. We assume that the platform of the rotary support, on which the antenna is installed, can perform these turns. In addition, we assume that the rotary support can move linearly in the direction of the normal to the scanning surface. The mechanisms by which these turns and movements are performed are not studied here. Theodolites, quadrants, measuring rulers, laser levels, laser range finders, laser trackers, etc., can be used as instruments for measuring angular and linear movements that are necessary for alignment. It is obvious that various methods of alignment can be implemented depending on the DUT and the available measuring equipment. For example, consider the simplest method of alignment of large antenna arrays proven in practice. Measuring instruments during the alignment were theodolite, quadrant, measuring rulers, and the mechanism of moving the probe across the scanning surface. Four measuring rulers oriented perpendicularly to the aperture plane and rigidly fixed at its corners were used as reference devices. Alignment of the antenna begins with the alignment of the rotary support platform before the DUT is installed onto it. The measuring tool for this operation is the quadrant, and the alignment tool is special adjusting screws. Then, the AUT is installed onto the platform and the leveling operation is repeated with the quadrant installed onto a special area that is included in the design of the DUT and determines the position of the horizontal plane of the object. With these operations, the desired roll position of the DUT is ensured. The next step is to set the aperture plane of the antenna to the desired position by rotating the antenna by azimuth and angle of elevation. The measuring instruments for performing these operations are theodolite, a measuring ruler fixed to the movable probe of the facility, and four reference devices made as measuring rulers oriented perpendicular to this probe and rigidly fixed at its corners. Alignment begins by selecting the location of the theodolite and setting the vertical sighting plane parallel to the plane of probe movement. The theodolite is installed on one side of the scan plane, and its location is chosen so that the measuring ruler fixed on the movable probe gets into the "field of view" of the theodolite when it is moved along the horizontal line (Fig. 4.37). Then, the azimuthal angle of the theodolite, at which its optical axis is parallel to the plane of probe movement, is determined. To do this, one selects two horizontal positions of the probe (with a measuring ruler attached to it) as reference points. The greater the distance between these points, the more accurately the above task will be solved. By means of the theodolite's optical system, record the readings of the ruler scale attached to the probe that fall on the

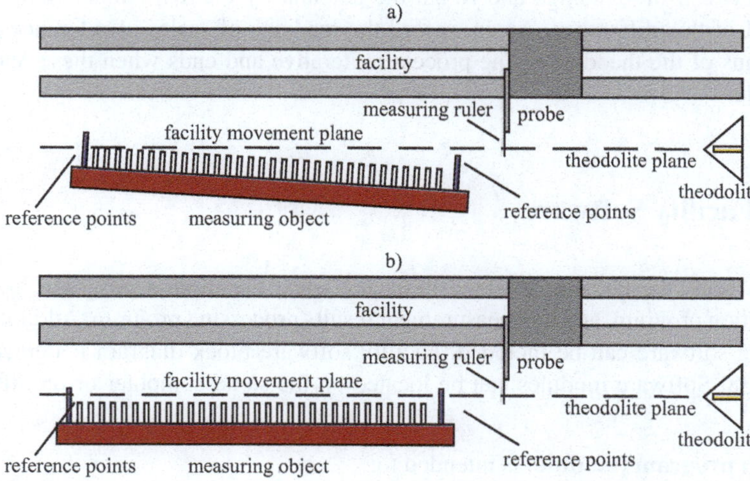

Fig. 4.37 Alignment of the DUT; **a**—position of the object before the alignment; **b**—position of the object after the alignment

optical axis of the theodolite at the reference points. If these readings are identical, the process of setting the azimuthal angle of the theodolite is complete. If these readings differ from each other, then turn the theodolite by azimuth by half the difference of these readings (remember that the theodolite displays the mirrored scale of the ruler). Repeat these operations until the readings of the measuring ruler measured with the theodolite at the reference points coincide. Fix the position of the theodolite by azimuth with a locking screw. Check whether the measuring rulers of reference devices get into the "field of view" of the theodolite and, if necessary, move the DUT to the desired side in the direction perpendicular to the scan plane. Figure 4.37 shows the alignment procedure for the DUT.

Perform alignment of the AUT by the angle of elevation using the theodolite and mechanism for moving the rotary support by the angle of elevation. Theodolite is moved only in the elevation plane. Using the optical system of the theodolite, record the readings of the scales of the two reference measuring rulers closest to the theodolite (fixed on the AUT), which get on the optical axis of the theodolite. If these readings are identical, the process of setting the angle of elevation of the AUT is complete. If these readings differ from each other, then determine the average difference between the readings and rotate the antenna by the angle of elevation to compensate for half of this difference. Again, record the readings of scales of reference rulers by means of the theodolite. The process is iterative and ends when these readings coincide. Perform alignment of the AUT by azimuth using the theodolite and mechanism for moving the rotary support by azimuth. Theodolite is moved only in the elevation plane. Using the optical system of the theodolite, record the readings of the scales of the reference measuring rulers closest and farthest to the theodolite. If these readings are identical, the process of aligning the AUT by azimuth is complete. If these readings differ from each other, then determine the average difference between the readings and rotate the antenna by the azimuth to compensate for half of this difference. Again, record the readings of scales of reference rulers by means of the theodolite. The process is iterative and ends when these readings coincide.

4.9 Facility Software

Facility software consists of three basic modules: the motion program, the data acquisition program, and the measurement results processing program. Additionally, aligning software can be included. Facility software block diagram is depicted in Fig. 4.38. Software modules can be located on the same computer or on different ones.

Motion program (module) is intended to:

- Input facility motion parameters (such as probe trajectory parameters, speed, and tower acceleration and deceleration parameters);
- Emit commands for special probe movements (such as positioning to given point).

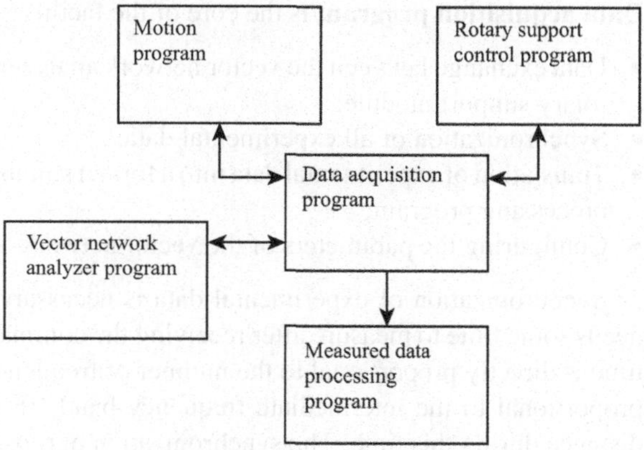

Fig. 4.38 Block diagram of the planar near-field facility software

Communication with the facility controllers is implemented with standard industrial interfaces such as RS-485, CAN, LAN, and GPIB. Embedded interface boards or external converters such as USB/CAN are used for that purpose. Depending on the system design, either data from linear encoders comes to application, which process it and emit correction commands for controllers or, if encoders connected to controllers, application exchanges data with controllers. Last variant is more reliable and quick operating, since in this case motion program only exchanges and visualizes data. In addition, this variant is less prone to failures due to operation system delays if Windows system is used. For multicore processors, one or several cores can be rigidly assigned for program execution. However, this leads to an overall decrease in computer performance. Program priority can be elevated, but this also decreases overall performance. If computer is used to process coordinates or receives data at each scan point (and not at the end of the line), it is recommended to use the Linux operating system. Using Windows operating system in such a situation does not allow to fully use the capabilities of the computer and, in particular, process the data of previous measurement, while next measurement is in progress. Motion program can be written in any programming language (C/C#, Java, and so on).

Rotary support control program (module) is intended for DUT orientation with rotary support drives. The specific functions performed by this program depend on the type of rotary support used. In case of fully automated rotary support, following commands are executed:

- Motion of rotary support tower along the horizontal guide-rails to set the required distance from the scan plane to the DUT;
- Rotation DUT by azimuth;
- Rotation DUT by elevation.

In addition, if controllers and rotary support allow, this module can be used to expand the functionality of the planar facility by giving it the ability to rotate the measurement object by 360°, which allow scanning the radiation field of the object on a cylindrical surface.

Data acquisition program is the core of the facility software. It is intended for:

- Data exchange between the vector network analyzer, the motion program, and the rotary support module;
- Synchronization of all experimental data;
- Translation of experimental data into a format suitable for the measurement results processing program;
- Configuring the parameters of the vector network analyzer.

Synchronization of experimental data is necessary because the vector analyzer needs some time to measure after receiving the command "start measurement." This time is directly proportional to the number of frequency sweep points and inversely proportional to the intermediate frequency band (IF band). Probe moves at some distance during this time. The synchronization purpose is to bind the amplitude and phase values of the complex transmission coefficient at each frequency to the real physical coordinate at which they were measured. The data acquisition program can act as a unified control wrapper, into which the motion program, the align module, and the vector network analyzer module are integrated.

Vector network analyzer program (module) provides:

- User access to the instrument settings (parameters to be configured are: S-parameter type, start and stop frequencies, number of sweep points, calibration file, intermediate frequency band (IF band)) from graphical window;
- Software trigger measurements;
- Access to data in the requested format (real and imaginary part of the complex number, amplitude in dB and phase, and so on).

This module is based on principles of open architecture to allow connecting with another device of the same manufacturer or with the devices of different manufacturers. The programming language of the module is determined depending on the used drivers of the hardware interface. The input/output (I/O) library is supplied by the device manufacturer and should be installed and registered on the control computer. The most common languages are: C, Microsoft Visual C++, Microsoft Visual Basic, MATLAB®, and LabWindows/CVI and LabVIEW test environments. ASCII sequences that are subject of instrument and controlling computer exchange are independent of the physical interface. The hardware structures used to establish the connection, configure it, and read the response of the device directly affect the internal content of the I/O library used by the control computer. Thus, the choice of the I/O library turns out to be significantly dependent on the hardware implementation of the physical interface. As a rule, devices of the same manufacturer have the same interface and different command sets. Interfaces of different manufacturers can differ from each other. Structure of control program utilizing VISA I/O library is depicted in Fig. 4.39.

The most promissory standard is LAN eXtension for Instrumentation (LXI) one, and most promissory interface for measuring device control is a Web browser.

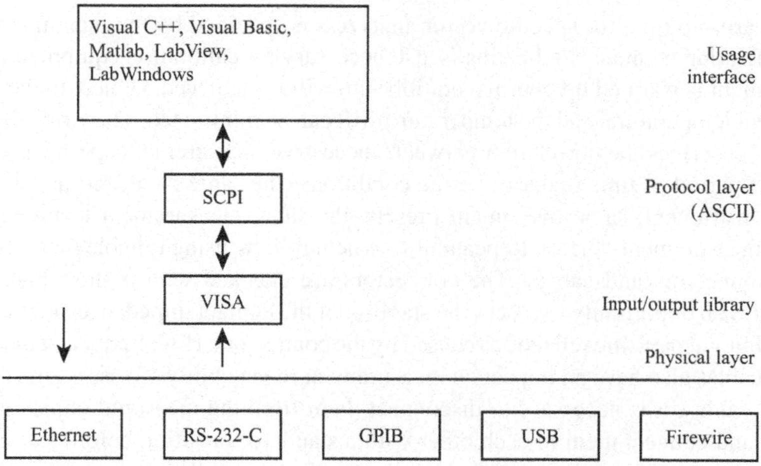

Fig. 4.39 VISA I/O library and SCPI standard

Measured data processing program performs the following functions:

- Calculation of probe-corrected 3D RP based on the data measured at the scan plane;
- Calculation of amplitude–phase distribution at any plane parallel to the scan plane, including the AUT aperture (i.e., especially relevant for the defectoscopy of different types of antenna arrays);
- Generation of 2D RP in any section of 3D RP;
- Calculation of antenna radio parameters, such as gain, directivity, and side lobe level (at specified angles or in a specified angular range);
- Visualization (graphical representation) of the results;
- Formation of measurement protocols and their passing to the printer, to flash drive or to LAN.

4.10 Instrumental Measurement Errors

The result of the measurement of any parameter differs from its true value by the measuring inaccuracy. There are two main types of these inaccuracies:

- Random measurement errors (they can only be described statistically but not corrected);
- Systematic measurement errors (such errors occur for known reasons and can be partially corrected by relevant computational procedures).

Random measurement errors include errors caused by temperature drift, repeatability errors, and errors caused by thermal noise. To reduce the impact of temperature drift, it is necessary to preheat the equipment, even if it has good temperature stability.

The warm-up time for specific vector analyzers is specified by the manufacturer in the operating manuals. Preheating is also necessary for calibration equipment. If the equipment is warmed up, thermal equilibrium will be achieved, subject to the stable ambient temperature, and the temperature drift can be minimized. The term "repeatability" describes the correlation between successive measurements performed over a short period of time under the same conditions (the same measured quantity, the same instrument, same instrument presets, the same measurement technique, the same measurement object). Repeatability is achieved by using reliable measurement port connectors and cables. The connectors are checked with a short load. This provides an opportunity to assess the stability of the contact impedance and the level of additional parasitic reflections caused by the connectors. High frequency measurement cables also have an impact on measurement repeatability. To assess the quality of the cables, it is necessary to disconnect them from the measured object and the probe and connect them to each other with a standard transition before starting the measurements. Then, the magnitude of the amplitude and the phase of the complex transmission coefficient of this connection have to be measured, and the result has to be remembered. In addition, the effect of cable deformation on measurements within the allowable bending radii of these cables has to be evaluated. The condition of the connectors and their proper maintenance also affect the repeatability of the results. Measuring connectors are subject to the following requirements:

- The contact surfaces of the connectors must be free from contamination (to clean the connectors, use cotton swabs which do not leave fibers on the surface, soaked in alcohol, and after wiping the connectors must be blown through with compressed air);
- Not to damage the thread when tightening the coupling nuts of the connectors (tightening must be done with a torque wrench, and to prevent damage to the copper core of the cable and unnecessary stress in the inner contacts of the connector, the tightening must be performed by holding the body of the connector in a fixed position, and only the nut must be turned).

Thermal noise superimposed on the useful signal leads to a random measurement error. At room temperature (290 K), the power density of the thermal noise is 4 $\times$ 10^{-21} W/Hz, which in logarithmic units corresponds to -174 dBm (it should be reminded that dBm is calculated as 10 lg of the ratio of the value measured to 1 mW). If we were dealing with an ideal device with no internal noise, and if the IF filter has an ideal rectangular characteristic, then only the noise of the specified value would be superimposed on the measured signal. However, any measuring device has its own noise, and the characteristic of its IF filter differs from the rectangular one, which leads to a decrease in the signal-to-noise ratio. Some increasing in this ratio can be achieved by using direct inputs of the receiver. There are two types of systematic errors: nonlinear errors and linear errors. Nonlinear errors include the compression effect, which is caused by the properties of the mixers. This effect is due to the difference in amplitudes in the measuring and reference channels. The range of levels in the reference channel is determined by the range of adjustment of the output power of the measuring channel. This effect is evident in the measurements

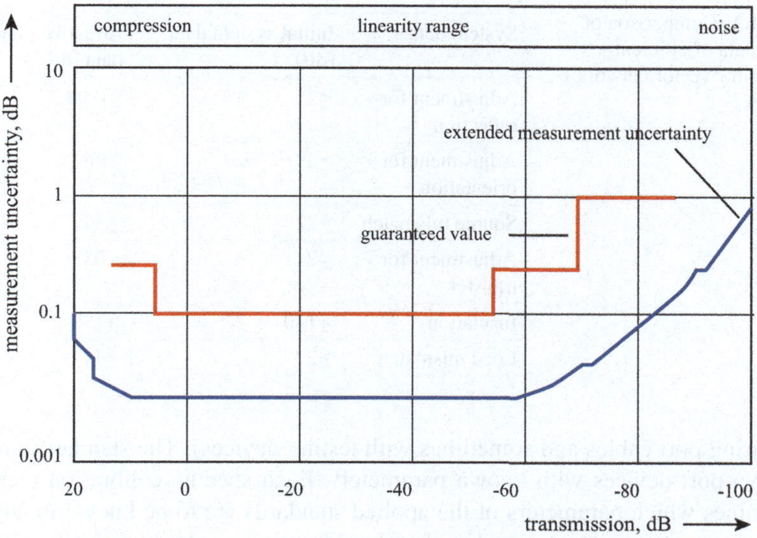

Fig. 4.40 U-shaped curve

on active devices with a high output power value. In this case, the receiver may be overloaded to compression mode. Typical measurement uncertainties that result from compression are shown in the left part of Fig. 4.40. Since the signal-to-noise ratio decreases at low signal levels, the linearity range may not be used for arbitrary low amplitude signals. Thus, for precision measurements, it is necessary to select a signal level that eliminates the compression effect described above, but at the same time provides a good signal-to-noise ratio. For measurements of the reflection coefficient and transmission coefficient of passive devices, the output level of the measuring port signal of the order of -10 dBm is usually a good compromise. When measurings are performed on devices with high amplification, it may be necessary to reduce the power of the signal source.

Vector analyzer can be represented in the form of distorting two-port circuit connected to an ideal network analyzer. The parameters of the distorting circuits can be designated as the components of the error. Most error components can be interpreted as source system data. The mathematical compensation of the distorting circuit effect on the measurement result is the correction of the system error. Systematic measurement errors that will remain uncompensated after system error correction can be interpreted as effective system data. They depend on the accuracy of the error component definition. The stability of the system error correction is limited by random measurement errors caused by temperature drifts, noise, etc. Table 4.2 shows a comparison of typical input data and the effective system data for a vector network analyzer.

The error components are determined with the use of the procedure known as calibration. Several preliminary measurements of the devices included in the calibration standard set are performed sequentially on the measuring unit (network analyzer with

Table 4.2 Comparison of source data of efficient systems for vector network analyzer

System data	Initial system data (d*B*)	Effective system data (d*B*)
Adjustment for reflection	≤2	≤0.04
Adjustment for orientation	≥29	≥46
Source mismatch	≥22	≥39
Adjustment for transfer	≤2	≤0.06
Insulation	≥130	≥130
Load mismatch	≥22	≥44

measuring port cables and sometimes with testing devices). The standards are one- and two-port devices with known parameters. Each specific calibration technique determines which parameters of the applied standards are to be known in advance. Since it is not possible to produce ideal calibration standards, specific values of internal deviations of standard are reported to the network analyzer (i.e., entered into it) in the form of characteristic data. Once the calibration procedure is completed, the network analyzer calculates the error components. During the calculation process, the analyzer uses the values it measures during the calibration and the characteristic data pertaining to the standards. Using the error components found, it is then possible to correct the measured values and calculate the corrected values of the *S*-parameters of the device. The physical interface between the virtual 2 N-pole of the error and the device under test is the so-called reference plane. It is to this plane that the measured and corrected *S*-parameters refer to. When coaxial calibration standards are used, the reference plane is set as the interface plane of the external conductor. Figure 4.41 shows the location of the reference plane.

Special one- and two-port devices are used as standards for calibration. The parameters of calibration devices are collected in sets according to the form of the characteristic data. This data is usually recorded in a digital format in the instrument's

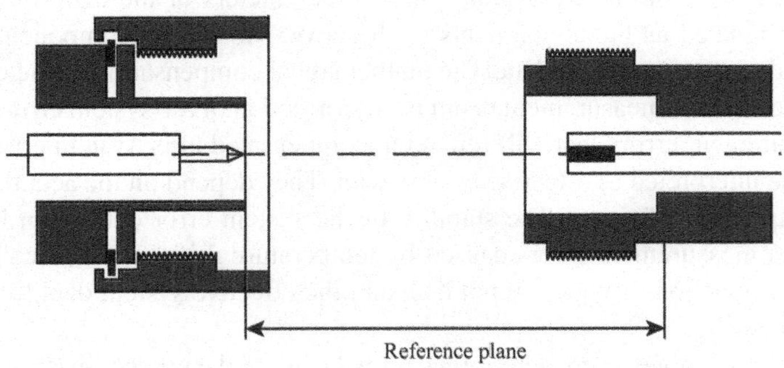

Fig. 4.41 Location of the reference plane in an N-type connector

memory. Standards can be described by means of special coefficients or by complex S-parameters. Description in the form of S-parameters requires an increase in the amount of data. The coaxial calibration kit includes:

- Coaxial short-circuit load;
- Coaxial open load;
- Match load;
- Sliding load;
- Reflection standard;
- Linear standard;
- The symmetric circuit standard;
- The attenuator standard;
- Unknown pass-through transfer standard.

In addition to coaxial calibration sets, there are waveguide and microstrip versions of calibration sets. To ensure the fastest and direct calibration process, most manufacturers of vector network analyzers offer equipment for automatic calibration. It allows the user to avoid the time-consuming and error-prone process of manual switching between different calibration standards. Since the automatic calibration equipment can store characteristic data internally, it is not necessary to transfer the data to a separate storage medium.

4.11 The Time Budget for Automated Measurements

The time taken for measuring an object at planar near-field facility depends on the following factors: the dimensions of the object; the operating frequency; the distances between the aperture and the scan plane; the speed of the testing facility movement; and the capacity of the vector analyzer used on the facility.

For high-quality calculation of the radiation pattern by the results of the measurement in the near field, it is necessary to intercept, if possible, the entire field emitted by the AUT. As it was mentioned above, the field level at the edge of the scan area should be 30–40 dB less than the field level at the maximum. The second way to determine the boundaries of the scanning zone is to calculate them based on the parameters of the object [4]. An example of such calculation is shown in Fig. 4.42.

As seen in Fig. 4.42, with large geometric dimensions of the object and high operating frequency, the size of the scan area (L) may become very big. The typical values of the probe speed at planar near-field facilities are from 100 to 300 mm per second. Thus, if the aperture of the object of measurement (D) is 1 m, and the work is performed at a frequency of 1,000 MHz, the scan size along the X-axis, at the distance from the scan plane to the aperture equal to 3λ and the angle θ_{max} equal to 60°, will be as follows

$$L = (1 + 2 \cdot 0.9 \cdot 1.72)\, m = 4.096\, m \tag{4.1}$$

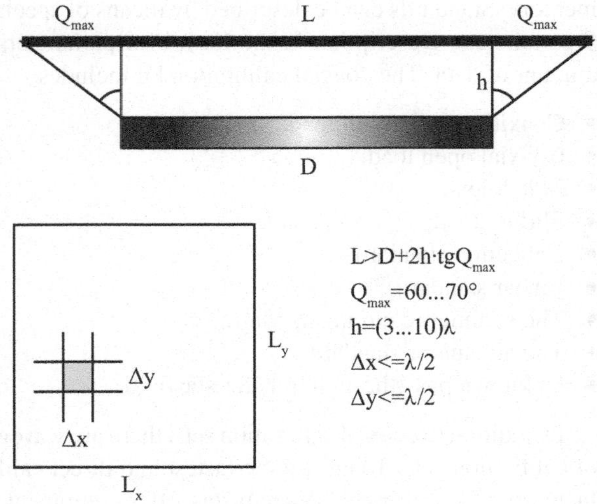

Fig. 4.42 Scan area size determination

and at the distance of 10λ, we have 11.32 m. At the speed of 150 mm/s (the area under consideration is the part with uniform motion, excluding the acceleration and deceleration parts), in the first case, the probe will pass this distance in 27 s, and in the second case, in 1 min 16 s. Assuming that the aperture of the measured object has the same height and width, then the size of the scan area in *y*-axis is equal to the scan size along the *x*-axis. Taking into account that for high accuracy of calculation of the radiation pattern, the step values on *x* and *y* must be less than λ/2, so we choose the discrete values on *y* equal to 100 mm. As a result, we have 41 lines for the first case and 113 lines for the second one. Thus, the total scan time in the first case will be 18 min 45 s, and in the second case, 2 h 22 min. Additionally, one can take into account the time required to move the probe carriage from line to line, but it does not make a significant contribution to the total time. Thus, at the same speed of 150 mm/s along the *y*-axis, the time will be 0.7 s per ascent, and since the number of lines is 41, the number of ascents is 41 minus one; therefore, the total time will be 31 s. From the example considered, it is clear that the distance from the aperture to the scan plane must be selected very carefully, with due consideration of many factors. To reduce the scanning time, the distance selected must be as small as possible. However, this is only true for operation at low frequencies, when the wavelength is large and provided that the probe does not exert a strong influence on the object of the measurement. As the frequency increases, it is necessary to move the scan plane away from the aperture. Let us see how the scanning time changes in our example with a tenfold change in frequency. Let us take the operating frequency equal to 10 GHz, with the wavelength of 3 cm. Then, at a distance from the aperture to the scan plane equal to 3λ, we have *L* equal to 1.31 m. With 10λ, it is 2.032 m. The passage time of the line will be 9 s in the first case and 14 s in the second case. However, in order to provide the required discrete on the *y* variable, it is necessary to change the number of lines. The discrete at this frequency will be 10 mm. Accordingly, the number of lines will be 131 in the first case and 203 in the second case. The total scan time will be 20 min in the first case and 47 min in the second case. Getting up the y will add 9 and 14 s, respectively. As it can be seen, with increasing frequency, both options became acceptable in terms of the time taken. The next factor to consider when calculating the time budget of the near-field facility is its speed. As noted above, in order to ensure the satisfactory accuracy of the radiation pattern calculation, it is necessary to provide the required steps on both x- and y-axes. Depending on the implemented kinematic scheme of the movement, uniform horizontal motion (scan) along x-axis and intermittent (stepped) vertical motion on y-axis or uniform vertical motion (scan) and intermittent (stepped) horizontal motion, respectively, it is necessary to provide the required discrete on x or on y (on scan axis). At the selected motion speed of 150 mm/s, the probe will travel a distance equal to the required step, in the first example (at an operating frequency of 1,000 MHz) in 0.8 s (a 100-mm step), and in the second example (at an operating frequency of 10 GHz) in 0.07 s (a 10-mm discrete). In the latter case, it is necessary to make sure that the vector analyzer used can perform the measurements within the available time. If the network analyzer does not have a buffer and the field values are transferred after each measurement, the type of interface used for communication between analyzer and the data acquisition PC affects the measurement time of one spatial point. Table 4.3 shows the measurement time of one sweep measured by the authors for various conditions. The measurements were carried out under the following conditions: calibration disabled, IF bandwidth is 1 kHz (digital sweep), frequency band from 3,000 to 12,000 MHz, 46 evenly distributed sweep points, trigger mode: single (the only mode used in near-field facilities). Time is given in ms. The measurements were carried out with the display output turned off and on.

An additional source contributing to the measurement time increase is the uneven movement of the tower within the working area. This is due to the large mass of the tower and its consequent inertia. The movement of the tower can be divided into three sections: acceleration, uniform movement, and deceleration. Figure 4.43 shows

Table 4.3 .Measurement time of different network analyzers.

Analyzer	D-Com, SDATA	SCPI, SDATA	
		Visible off	Visible on
Keysight PNA-X N5241A 4 ports	61	68	78
Agilent PNA-L N5230C 4 ports	79	84	104
R&S, ZVA (2017)		50	50
R&S, ZNB (2017)		48	48
R&S, ZVL (portable, 2017)		219	

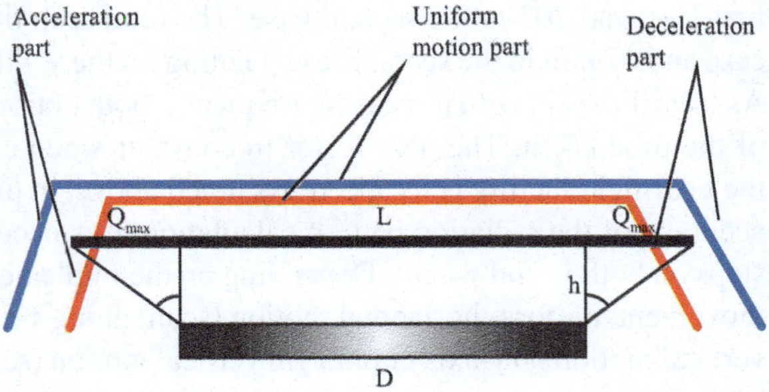

Fig. 4.43 Trajectory of the tower movement

the trajectory of the tower. In the areas of acceleration and deceleration, the tower of the facility gains speed and loses it, respectively. Consequently, the travel time in these sections increases, resulting in an overall increase in the measurement time. The speed in these sections normally increases gradually, which leads to uneven distribution of measurement points in the line.

In order to avoid this unevenness, the trajectory is expanded in the way that the acceleration and deceleration areas are outside the scan area, which, in turn, also contributes to the increased time required for the measurements.

Acknowledgements The authors express their gratitude to Smitek company for the provision of graphic and reference materials (https://smitek-robotics.com/en/wp-content/uploads/sites/2/2019/01/2018-06-smitek-catalog-eng.pdf).

Office address Skolkovo, Moscow, 121205, Russia.

Production facilities address 42, Volgogradsky Prospekt, Moscow, Russia, 109316, Technopolis «Moscow». **+7-495-2215143**

soft@smitek-robotics.com

References

1. Tailor-Made Antenna Measrument Systems, SMITEK, 2019
2. "SKBIS Ltd" web-site https://www.skbis.ru.
3. LM10D05_01. RLS: Rotary and Linear Motion Sensors.
4. Methods for measuring the radio characteristics of antennas in the time and frequency domain. Materials of the "Trim measuring systems" seminar, 2010

Chapter 5
Spherical Near-Field Antenna Measurements

5.1 Problem Statement, Basic Coordinate Systems

In Fig. 5.1, coordinate systems in use are shown: Cartesian with variables X, Y, Z and spherical with variables r, θ, φ.

Let us introduce the following notations for the kinematic scheme: roll (φ) and azimuth (θ).

5.2 Selecting the Size of the Scan Area and the Radius of the Measurement Sphere

The angular dimensions of the scan area are determined from considerations of geometric optics similar to the planar case. A distinctive feature of spherical scanning is the ability to provide a fully enclosed measurement surface. Therefore, it is reasonably to measure field on the entire sphere if it is mechanically possible. The near-field method may not be used at any distance; it requires the absence of reactive fields near the probe. One of the main factors when choosing a distance in favor of increasing it is to reduce the influence of the radiation pattern of the probe. If some antennas, in particular traveling wave antennas, are measured with a small scan radius, multiple reflections between the antenna and the probe can occur, creating a characteristic false dip at the zero azimuth direction. A factor to reduce the distance may be the lack of an adequate amount of radio absorbing material in partially anechoic chambers.

V. S. Kalashnikov et al., *Near-Field Antenna Measurements*, Springer Aerospace Technology, https://doi.org/10.1007/978-981-33-6436-3_5

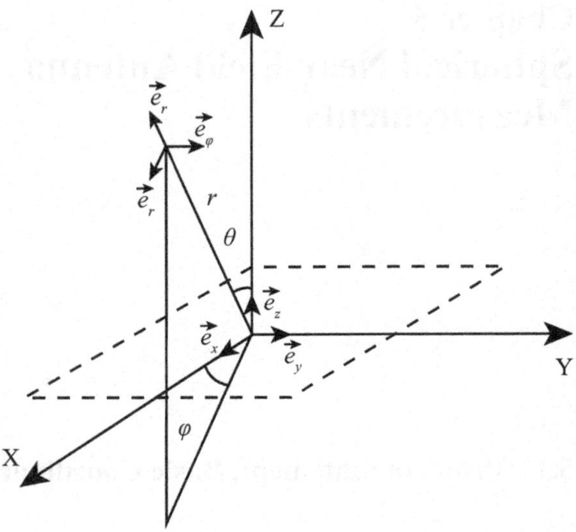

Fig. 5.1 Coordinate systems in use

5.3 The Algorithm to Calculate RP Based on the Measurement Results

In a linear, homogenous, isotropic medium, an electromagnetic field harmonically dependent on time satisfies the Maxwell equations in the following form:

$$\nabla \times \boldsymbol{H} = i\omega\varepsilon \boldsymbol{E} + \boldsymbol{J} \tag{5.1}$$

$$\nabla \times \boldsymbol{E} = -i\omega\mu \boldsymbol{H}$$

where $\boldsymbol{H}$ is the magnetic field strength vector; $\boldsymbol{E}$ is the electric field strength vector; $\boldsymbol{J}$ is the vector of electric current density; i is the imaginary unit; ω is the circular frequency; ε and μ—absolute dielectric permittivity and magnetic permeability of the medium, accordingly.

In source-free region ($\boldsymbol{J} = 0$), both vectors $\boldsymbol{E}$ and $\boldsymbol{H}$ satisfy the vector wave equation:

$$\nabla \times (\nabla \times \boldsymbol{F}) - k^2 \boldsymbol{F} = 0, \tag{5.2}$$

where $k = 2\pi/\lambda$ is the wave number.

The solution to the equation can be obtained by separating the variables in only six coordinate systems in which the Lamé coefficients satisfy certain relations. In practice, three of these systems are used [1, 2]. In the spherical coordinate system, the components of the electric field are expressed as a linear combination of vector spherical waves (harmonics) with coefficients a_{nm}, b_{nm}.

$$Er = \frac{k}{\sqrt{\eta}} \cdot \frac{1}{\sqrt{2\pi}} \sum_{n=1}^{\infty} \frac{1}{\sqrt{n(n+1)}} \sum_{m=-n}^{n} \left(-\frac{m}{|m|}\right)^{m} b_{nm} \cdot \frac{n(n+1)}{kr} h_n^{(2)}(kr) \overline{P}_n^{m}(\cos\vartheta) e^{im\varphi}, \tag{5.3}$$

$$E_{\vartheta} = \frac{k}{\sqrt{\eta}} \cdot \frac{1}{\sqrt{2\pi}} \sum_{n=1}^{\infty} \frac{1}{\sqrt{n(n+1)}} \cdot \sum_{m=-n}^{n} \left(-\frac{m}{|m|}\right)^{m} a_{\mathrm{nm}} h_n^{(2)}(\mathrm{kr}) \frac{im\overline{P}_n^{m}(\cos\vartheta)}{\sin\vartheta}$$
$$+ b_{\mathrm{nm}} \frac{1}{\mathrm{kr}} \frac{\mathrm{d}}{\mathrm{d(kr)}} (\mathrm{kr} \cdot h_n^{(2)}(\mathrm{kr})) \frac{\mathrm{d}\overline{\mathrm{P}}_{\mathrm{n}}^{|\mathrm{m}|}(\cos\vartheta)}{\mathrm{d}\vartheta} \mathrm{e}^{\mathrm{i}m\varphi}$$

$$E_{\varphi} = \frac{k}{\sqrt{\eta}} \cdot \frac{1}{\sqrt{2\pi}} \sum_{n=1}^{\infty} \frac{1}{\sqrt{n(n+1)}} \cdot \sum_{m=-n}^{n} \left(-\frac{m}{|m|}\right)^{m} a_{\mathrm{nm}} h_n^{(2)}(\mathrm{kr}) \frac{\mathrm{d}\overline{P}_n^{m}(\cos\vartheta)}{\mathrm{d}\vartheta}$$
$$+ b_{\mathrm{nm}} \frac{1}{\mathrm{kr}} \frac{\mathrm{d}}{\mathrm{d(kr)}} (\mathrm{kr} \cdot h_n^{(2)}(\mathrm{kr})) \frac{\mathrm{im}\overline{P}_{\mathrm{n}}^{|\mathrm{m}|}(\cos\vartheta)}{\sin\vartheta} \mathrm{e}^{\mathrm{i}m\varphi},$$

where $h_{\mathrm{n}}^{(2)}(\mathrm{kr}) = \sqrt{\frac{\pi}{2\mathrm{kr}}} \mathrm{H}_{\mathrm{n}+\frac{1}{2}}^{(2)}(\mathrm{kr})$ is the spherical Hankel functions;
$H_{n+\frac{1}{2}}^{(2)}(kr)$ is the Hankel function of the second kind with half-integer index;
$\eta = \sqrt{\varepsilon/\mu}$ is the wave impedance;
$\overline{P}_n^{|m|}(\cos\vartheta)$ are normalized associated Legendre polynomials:

$$\overline{P}_n^{m}(\cos\vartheta) = \sqrt{\frac{2n+1}{2} \frac{(n-m)!}{(n+m)!}} P_n^{m}(\cos\vartheta). \tag{5.4}$$

which, in turn, are related with Legendre polynomials:

$$P_n^{m}(\cos\vartheta) = (\sin\vartheta)^{m} \frac{\mathrm{d}^{m} \mathrm{P}_n(\cos\vartheta)}{\mathrm{d}(\cos\vartheta)^{m}}, \tag{5.5}$$

$$P_n(\cos\vartheta) = \frac{1}{2^n n!} \frac{\mathrm{d}^n}{\mathrm{d}(\cos\vartheta)^{\mathrm{n}}} (\cos^2\vartheta - 1)^n. \tag{5.6}$$

The magnetic field components are expressed in a similar way. Based on the orthogonality property of the vector spherical waves, the relationship between the coefficients a_{nm} and b_{nm} and the field values on the scan sphere can be obtained (similar to the derivation of the coefficients of the Fourier series):

$$a_{\mathrm{nm}} = \frac{\sqrt{\eta}}{k} (-1)^{m} \frac{1}{h_n^{(2)}(\mathrm{kr_0})} \frac{1}{\sqrt{2\pi}} \frac{1}{\sqrt{n(n+1)}} \left(\frac{m}{|m|}\right)^{-m}$$

$$\times \int_{\theta=0}^{\pi}\int_{\varphi=0}^{2\pi} E_{\vartheta}(r_0,\vartheta,\varphi)\frac{i(-m)\overline{P}_n^m(\cos\vartheta)}{\sin\vartheta}$$

$$- E_{\varphi}(r_0,\vartheta,\varphi)\frac{\mathrm{d}\overline{P}_n^m(\cos\vartheta)}{\mathrm{d}\vartheta}\mathrm{e}^{-\mathrm{i}m\varphi}\sin\vartheta\,\mathrm{d}\vartheta\,\mathrm{d}\varphi \tag{5.7}$$

$$b_{nm} = \frac{\sqrt{\eta}}{k}(-1)^m \frac{1}{\left(\frac{1}{\mathrm{kr}}\frac{\mathrm{d}}{\mathrm{d(kr)}}(\mathrm{kr}\cdot h_n^2(\mathrm{kr}))\right)_{r=r_0}}\frac{1}{\sqrt{2\pi}}\frac{1}{\sqrt{n(n+1)}}\left(\frac{m}{|m|}\right)^{-m}$$

$$\int_{\theta=0}^{\pi}\int_{\varphi=0}^{2\pi}\left(\mathrm{E}_{\vartheta}(\mathrm{r}_0,\vartheta,\varphi)\frac{\mathrm{d}\overline{\mathrm{P}}_\mathrm{n}^{|\mathrm{m}|}(\cos\vartheta)}{\mathrm{d}\vartheta}+\mathrm{E}_{\varphi}(\mathrm{r}_0,\vartheta,\varphi)\frac{\mathrm{i}(-\mathrm{m})\overline{\mathrm{P}}_\mathrm{n}^{|\mathrm{m}|}(\cos\vartheta)}{\sin\vartheta}\right)\mathrm{e}^{-\mathrm{i}m\varphi}\sin\vartheta\,\mathrm{d}\vartheta\,\mathrm{d}\varphi$$

where r_0 is the scan radius (the distance from the intersection point of the rotation axis of the antenna rotary support to the phase center of the probe).

There is also an equivalent notation with $\sin(m\varphi)$ and $cos(m\varphi)$ instead of $\exp(-im\varphi)$ and $\exp(im\varphi)$. In this notation, four coefficients instead of two are required [2]. Thus, by measuring the values of the components of the field $\mathbf{E}_\theta$ and on a certain sphere, one can determine $\boldsymbol{E}_\theta$, $\boldsymbol{E}_\varphi$ and $\boldsymbol{E}_r$ on any outer sphere. In the roll-over-azimuth kinematic scheme, $\boldsymbol{E}_\theta$ is measured at the horizontal polarization of the probe, and the value of $\boldsymbol{E}_\varphi$ at vertical (such that the probe rotates relative to the horizontal polarization 90° clockwise when viewed from the AUT). To obtain field on a full sphere, it is enough to change azimuth angle from 0 to 180° (for the conditions described above—clockwise for a top view). It is important to note that for the correct roll (φ) measurement in the scheme with movable AUT, the latter should rotate clockwise, as seen from the probe. In this case, in the coordinate system associated with the AUT, the probe positions correspond to the standard spherical coordinate system. A required number of vector spherical waves in expressions (5.3) depends on the radius of the minimum sphere centered at the coordinate system origin, encompassing the entire AUT:

$$N = [kr_a] + n_1 \tag{5.8}$$

The parameter n_1, as a rule, is chosen equal to ten. The smaller the number N, the less the influence of reflections on the measurement result. The number of harmonics determines the minimum required number of points, and, therefore, the step between them along two coordinates ($\Delta\theta$, $\Delta\varphi$) (based on sampling theorem):

$$\Delta\theta = \Delta\varphi = 2\pi/(2N+1) \tag{5.9}$$

The step and the number of harmonics may differ for two coordinates, then instead of r_a, the semi-axes of the ellipsoid of revolution are used, this requires additional a priori knowledge of the antenna, but allows one to obtain a more accurate pattern due to the filtering property of the generalized Fourier transform. In practice, it may be useful to increase the sampling rate (decrease step), as this averages and reduces random errors. Also, a smaller step helps to filter out the non-propagating part of the field when scanning in the reactive region [3]. However, on the other hand, the limiting factor for the step reducing is the total measurement time.

Let us now consider some practical aspects of calculation using the computing of the integral for directivity with Octave or MATLAB language as an example.

- The measured data arrays have finite sizes; therefore, integration is replaced by (usually method of rectangles) summation.
- positioners angles are expressed in degrees, not in radians.
- A feature of the Octave language is the widespread use of matrix notation instead of loops, which significantly speeds up calculations due to the internal implementation with SIMD instructions.

According to all the above, the calculation of the full sphere directivity integral:

$$\int_0^{\pi}\int_0^{2\pi} P(\vartheta,\varphi)\sin\vartheta \, d\vartheta \, d\varphi \tag{5.10}$$

can be carried out in the following equivalent ways:

```
1.      SindThFull=ones(Numel(Ph),1)*sind(Th);
2.      Int=sum(sum(P.*sindThFull))*dTh*dPh*pi^2/(360*180);
```

Or in a new version of the language with autoexpansion of dimensions, which automatically expands `sind(Th)` from size $[1, L_\theta]$ to size $[L_\varphi, L_\theta]$ before multiplying with the `P`:

```
1.      Int=sum(sum(P.*sind(Th)))*dTh*dPh*pi^2/(360*180);
2.      Int=sum(P.*sind(Th)).')*dTh*dPh*pi^2/(360*180);
```

where P is the matrix of size $[L_\varphi, L_\theta]$; Th is the vector of azimuth angles expressed in degrees of size $[1, L_\theta]$; Ph is the vector of roll angles expressed in degrees of size $[L_\varphi, 1]$; dTh is the θ step, scalar; dPh is the φ step, scalar.

Both variants are much faster than using loops. Better one can be chosen based on the runtime (measured using the `tic` and `toc` functions), as well as on the RAM consumption.

The flowchart of the algorithm for calculating antenna far field (RP) based on experimentally measured amplitude–phase distribution at the spherical surface is shown in Fig. 5.2.

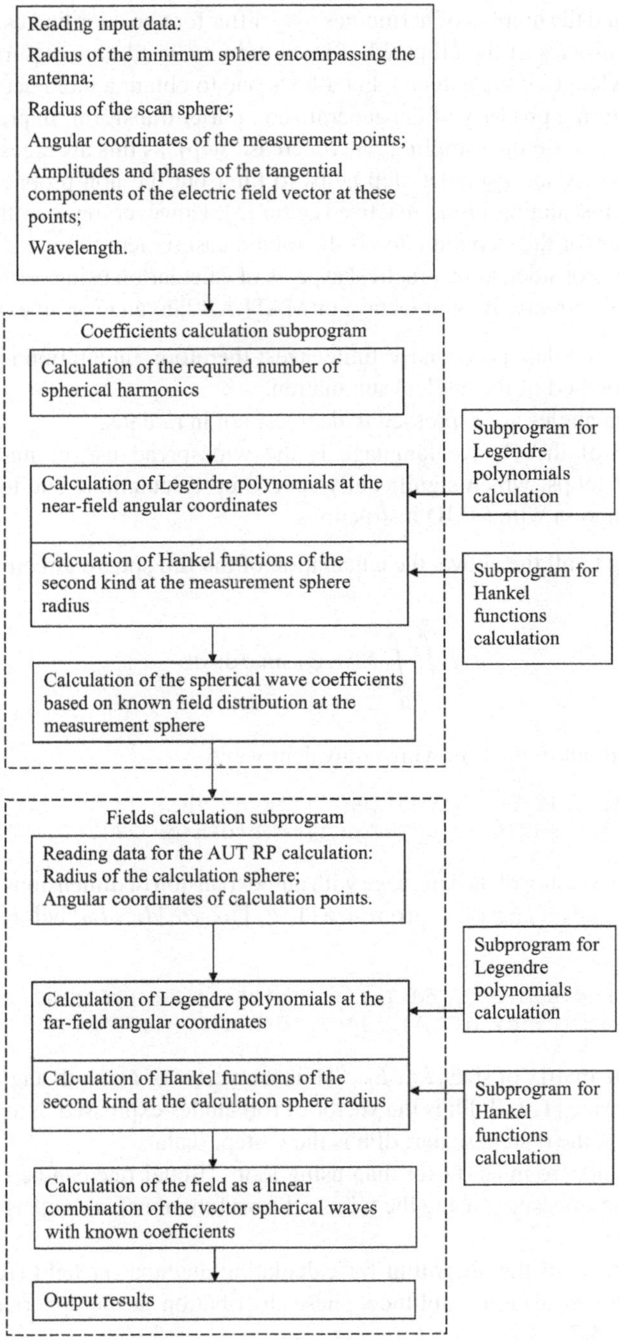

Fig. 5.2 Flowchart of calculating antenna far field (RP) based on the measured amplitude–phase distribution on the spherical scan surface

5.4 Probe Correction

As probes for measurements on a sphere, the open-ended rectangular waveguide (since it has an analytical solution), a circular waveguide with a TE11 wave or a vibrator are traditionally used, since they have a wide RP close to that for an ideal probe—a short dipole [1]. A typical AUT RP distortion at spherical measurement is widening in one plane and narrowing in another. Spherical near-field measurements are least affected by the probe RP [4]. If probe RP is wide, probe correction is often omitted. It has been accepted that the influence of the probe can be neglected if its RP changes by no more than 1 dB within the angles at which the antenna is visible from it. The last condition can be one of the main factors in choosing the radius of the measurement sphere. Expressions for correction of a single-mode probe, i.e., such that it is completely characterized by radiation patterns in two perpendicular sections, are given in [1]. An interpretation of the procedure for the multimode probe is given in [5]. Pockels and Faraday effect probes are increasingly used. In this type of probes, the received electric field changes optical fiber properties. The signal passing through fiber interferes with reference one, and, thus, information unambiguously associated with the required field is registered. Such a probes can be metal-free [6], which reduces the reflections between the antenna and the probe. Also, those probes have a small size and omnidirectional RP, which further reduces the need for probe correction. An additional advantage is the ultra-wide bandwidth of the probe, for example, in the range from 1 to 10 GHz [7]. There are also materials that allow reaching the 150 GHz band for Vivaldi antennas. These materials are listed in [8]. The use of optical fiber probes improves signal to noise ratio and phase stability. The disadvantages of the described design include the inability to work at high powers [3].

5.5 Algorithms for Calculating Antenna Directivity and Gain Based on Measurement Results

Since in spherical case measurement is carried out on the closed surface, directivity in direction (ϑ_0, φ_0) is calculated by definition as the ratio of the power radiated in the given direction to all radiated power.

$$D = \frac{4\pi P(\vartheta_0, \varphi_0)}{\int_0^{\pi}\int_0^{2\pi} P(\vartheta, \varphi)\sin\vartheta \mathrm{d}\vartheta \mathrm{d}\varphi} \tag{5.11}$$

The calculation of the gain is performed similarly to that for planar measurements with the difference that the far field is calculated with spherical near-field to far-field transform. When three antenna measurements are required, it is reasonable to use spherical facility. This is due to the fact that it is possible to intercept all the energy of the antenna with any RP. In addition, during measurements on the full sphere, the Gibbs effect caused by a field discontinuity at the edge of the scanning surface does not arise. Thus, there are no RP oscillations, and no error in calculated gain caused by them. The latter is especially true for low-frequency antennas, for which it is often difficult to achieve a decrease in the field at the boundaries of the scan area to −20 dB relative to the maximum of the amplitude distribution.

5.6 Algorithm for Calculating Antenna Aperture Field Based on Spherical Amplitude–Phase Measurements

Calculation of the amplitude–phase distribution in the AUT aperture is important for the diagnostic purposes. This allows one to identify faulty antenna array elements or "current surfaces" defects. Direct measurements near the surface of interest would cause a large error due to the mutual influence of the probe and antenna. The aperture diagnostic is widely used for planar measurements because the scan surface and the surface of interest have same shape, and thus, the calculation is relatively easy. If there is a priori knowledge about the antenna, for example, it is an equidistant array, then methods that require a smaller scan area are used. For example, scanning on one circle for linear antenna arrays, which called the "method of the meridian RP in the Fresnel region" [9]. RP of the single element of the antenna needs to be known for this method [10]. This method is relevant for use in anechoic chambers; circle scanning can be implemented on a spherical or cylindrical measurement facility. Returning to the amplitude–phase method, a number of methods for reconstructing the field in the aperture are used in practice, for example, by calculating the currents of equivalent electric or magnetic dipoles [11, 12]. Calculation using the far-field approximation is also used [2]. Consider this method.

In the far-field region, electric field strength $\mathbf{E}(\mathbf{r})$ and the coefficients of the plane wave spectrum $\mathbf{A}(\mathbf{k})$ are related as follows:

$$\boldsymbol{E}(r) = i\frac{2\pi}{r}k_2 A(k)\exp(-ikr) \tag{5.12}$$

Cartesian projections of the $\mathbf{A}$ are expressed as follows:

$$\boldsymbol{A}_x = \frac{\boldsymbol{E}_x}{i\frac{2\pi}{r}k_z\exp(-ikr)} \tag{5.13}$$

$$\boldsymbol{A}_y = \frac{\boldsymbol{E}_y}{i\frac{2\pi}{r}k_z\exp(-ikr)}$$

The $\boldsymbol{E}_x$ and $\boldsymbol{E}_y$ components, in turn, are related by known expressions with the spherical components $\mathbf{E}_\theta$ and $\mathbf{E}_\varphi$, whence:

$$\boldsymbol{A}_x = \frac{\cos\theta\cos\varphi\boldsymbol{E}_\theta - \sin\varphi\boldsymbol{E}_\varphi}{i\frac{2\pi}{r}k_z\exp(-ikr)}$$

$$\boldsymbol{A}_y = \frac{\cos\theta\sin\varphi E_\theta + \cos\varphi\boldsymbol{E}_\varphi}{i\frac{2\pi}{r}k_z\exp(-ikr)} \tag{5.14}$$

However, the above expressions to the left and to the right of the equal sign are written for different geometries. To overcome this difficulty, it is necessary to eliminate the dependence on r. It is proposed here to use the fact that, for sufficiently large values of the argument, the Hankel functions in the expressions for the components of E asymptotically tends to the Green function $\exp(-ikr)/r$:

$$H_\nu^{(2)}(x) = \sqrt{\frac{2}{\pi x}}\exp[-i(x - \frac{\pi}{2}\nu - \frac{\pi}{4})] + O(x^{-3/2}) \tag{5.15}$$

$$\frac{1}{\sqrt{r}}H_{n+1/2}^{(2)}(\text{kr}) \approx \frac{1}{\sqrt{r}}\sqrt{\frac{2}{\pi\text{kr}}}\exp(-ikr)\exp[i(\frac{\pi}{2}(n+\frac{1}{2})+\frac{\pi}{4})]$$
$$= \frac{1}{r}\sqrt{\frac{2}{\pi k}}\exp(-ikr)\exp[i(n+1)\frac{\pi}{2}] \tag{5.16}$$

Let us carry out similar procedure for $\frac{1}{r}\frac{\partial}{\partial r}(\sqrt{r}H_{n+1/2}^{(2)}(kr))$.

$$\frac{1}{r}\frac{\partial}{\partial r}(\sqrt{r}H_{n+1/2}^{(2)}(kr)) \approx \frac{1}{r}\frac{\partial}{\partial r}(\sqrt{r}\sqrt{\frac{2}{\pi kr}}\exp[-i(kr - \frac{\pi}{2}(n+\frac{1}{2}) - \frac{\pi}{4})]) =$$
$$\frac{1}{r}\sqrt{\frac{2}{\pi k}}(-ik)\exp(i\frac{2\pi}{4})\exp(-ikr)\exp(in\frac{\pi}{2}) = \sqrt{\frac{2k}{\pi}}\frac{1}{r}\exp(-ikr)\exp(in\frac{\pi}{2}) \tag{5.17}$$

Thus:

$$\frac{\frac{1}{\sqrt{r}}H_{n+1/2}^{(2)}(kr)}{i\frac{2\pi}{r}k_z\exp(-ikr)} \approx \sqrt{\frac{2}{\pi k}}\frac{\exp[i(n+1)\frac{\pi}{2}]}{2\pi i k_z}$$
$$\frac{\frac{1}{r}\frac{\partial}{\partial r}(\sqrt{r}H_{n+1/2}^{(2)}(kr))}{i\frac{2\pi}{r}k_z\exp(-ikr)} \approx \sqrt{\frac{2k}{\pi}}\frac{\exp[in\frac{\pi}{2}]}{2\pi i k_z} \tag{5.18}$$

One can check coincidence of units in both parts of each expression (k has unit 1/m). Now let us express the angles θ and φ of the spherical coordinate system through the Cartesian projections of the wave vector $\mathbf{k}$:

$$\begin{aligned} \sin\theta &= \sqrt{k_x^2 + k_y^2} \\ \cos\theta &= \sqrt{1 - k_x^2 - k_y^2} \\ \varphi &= \arctan(k_x/k_y) \end{aligned} \tag{5.19}$$

Substituting expression (5.3) and (5.18, 5.19) in succession into expression (5.14), one can obtain expressions for the PWS with which the field in the aperture is calculated (see Chap. 3). The fundamental restriction of the method due to the transition to plane waves is that only the field in the front hemisphere is taken into account. It should be noted:

- arctan function should be defined in a way to has range of values $(-\pi, \pi]$, not $[-\pi/2, \pi/2]$. Such a function is called atan2 in many programming languages.
- During calculations, first equidistant k_x, k_y mesh should be defined, and then, θ and φ angles should be calculated from it.

The results of applying the described algorithm to measurements of a linear antenna array with known excitation coefficients of radiators are shown in Fig. 5.3a, b. The measurements were carried out for the main polarization. In Fig. 5.3a, all radiators operate normally, and in Fig. 5.3b, one radiator is covered with radio absorbing material, which simulates its fault.

Figure 5.4 shows the results of applying the described algorithm to measurements of the R&S HF907 antenna (broadband Vivaldi radiator enclosed in a horn) for calculating the cross-polarization field component in the antenna aperture.

5.7 Errors in Spherical Measurements

The amplitude–phase method has an important feature that increases the accuracy of measurements. Since the data is processed by a Fourier-like algorithm, pips are filtered and random errors arising from point to point are smoothed out. The list of measurement errors from 22 items that affect the calculated RP for the cylindrical case is given in [13]. Expressions and references to methods for determining the error of the final result (RP, gain) from temperature, mismatch, polarization alignment are given for the far-field region in [14, 15]. Let us discuss results of numerical modeling the influence of random errors on the measurement results using the Monte Carlo method. This method allows to establish the errors of end result using a numerical experiment. This is done as follows. For each input variable, the distribution of random errors to which it is exposed is specified. Then, a set of "attempts" are made, in each of which all input variables are exposed by certain errors and output variables are calculated. Thus, for the output values, a set of values is obtained for which the mean and standard deviation can be found. The method allows avoiding bulky analytical calculations. An important feature that allows to reduce the computation time is the independence of the input data for individual attempts. Therefore, the

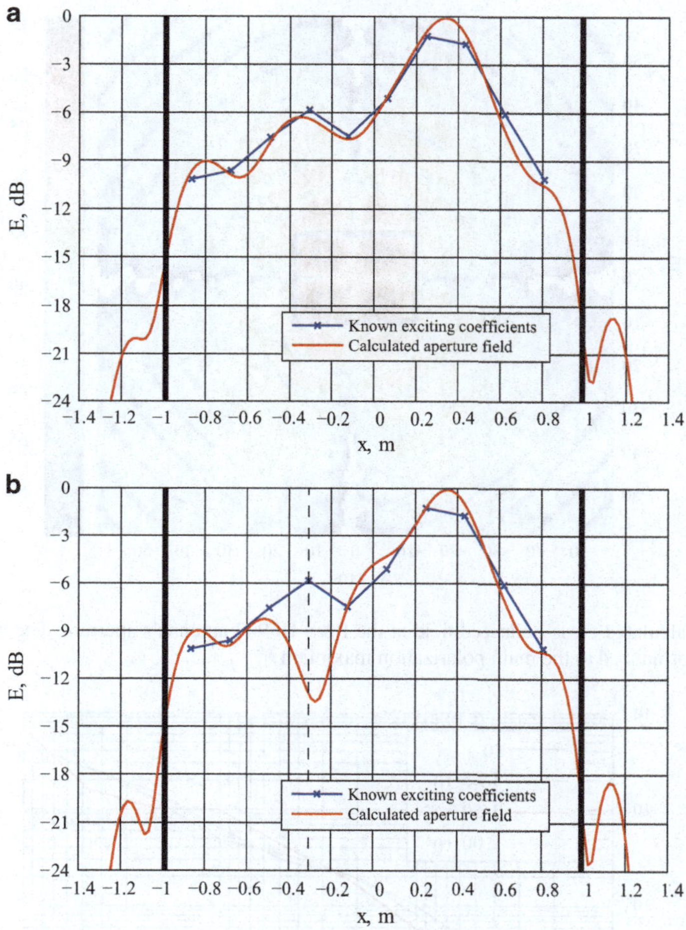

Fig. 5.3 **a** Calculated field of main polarization at the aperture of the linear antenna array with known exciting coefficients. All radiators operates normally. **b** Calculated field of main polarization at the aperture of the linear antenna array with known exciting coefficients. One of the radiators is covered with absorber, which simulates its fault

calculations can be parallelized in a computer network. For the spherical near-field to far-field algorithm, the input parameters are: angles θ and φ, radius R; tangential electric field components $\boldsymbol{E}_\theta$ and $\boldsymbol{E}_\varphi$. Output variables are $\boldsymbol{E}_\theta$ and $\boldsymbol{E}_\varphi$ at the far-field sphere. Figure 5.5 shows the results of a numerical experiment to study the effect of measurement sphere positioning errors on the far-field $\boldsymbol{E}_\theta$ and $\boldsymbol{E}_\varphi$ values. The antenna array of 8 × 10 elements with the half-wave element distance and constant amplitude and phase distribution was considered. Smallest scan radius is $r_0 = 7{,}68\lambda$. Normally distributed random noise with the standard deviation σ (horizontal axis) was added to the equidistant angular coordinates. The field on the input sphere generated for an ideal and distorted grid of angles was generated using an auxiliary

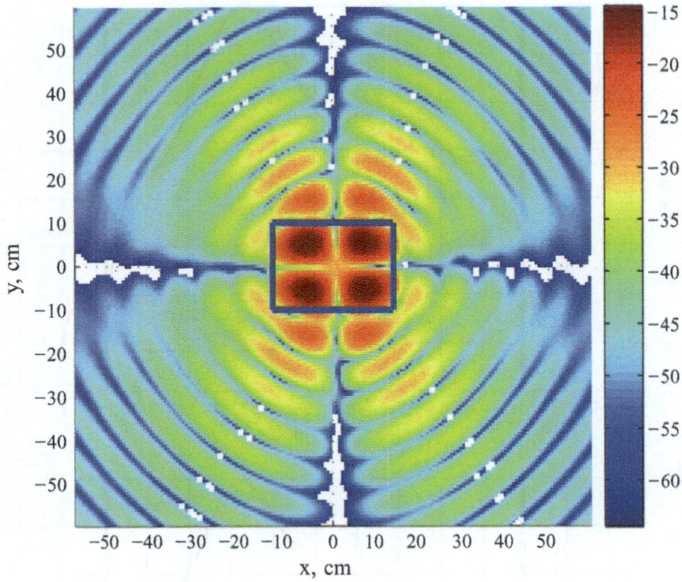

Fig. 5.4 Calculated cross-polarized field at the R&S HF907 antenna's aperture. The field values in dB are normalized to the main polarization maximum

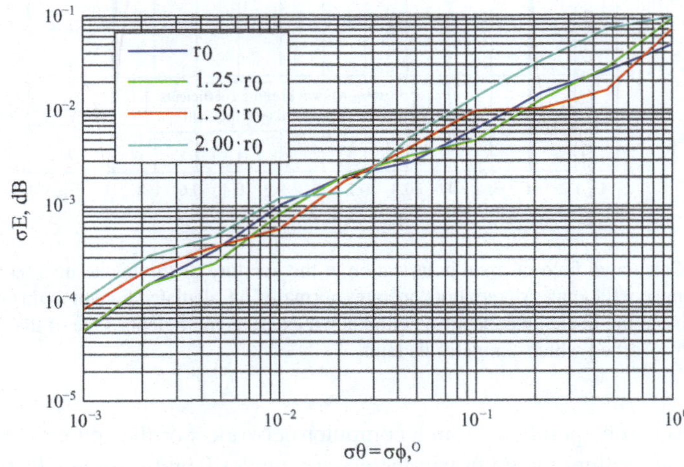

Fig. 5.5 Dependence of the RP standard deviation on the standard deviation of the normally distributed noise of the angular coordinates

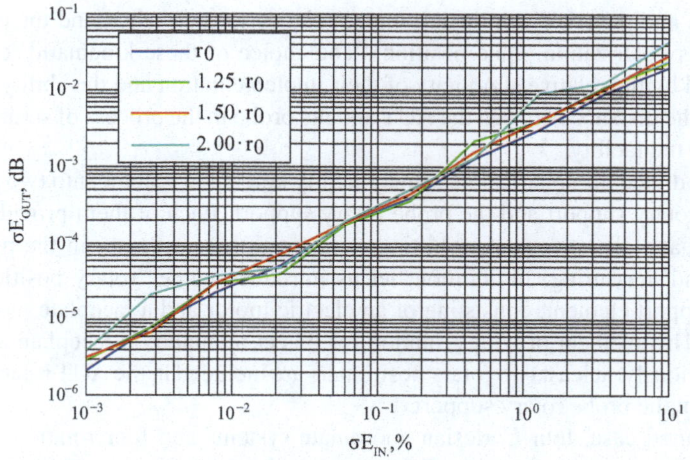

Fig. 5.6 Dependence of the RP standard deviation on the standard deviation of the normally distributed noise of the measured field strength

program. This program calculates the field of the dipole antenna array as sum of the fields of the elements; the field of each element is calculated with known rigorous analytical formulas. Then, the near-filed to far-field transform was performed. Then, the standard deviation between the obtained RP and the noise-free RP was calculated. The calculations was performed for angles $\varphi = 0°, \theta < 60°$. All calculations for each standard deviation value was performed 10 times (a new noise realization was generated each time) and averaged. Also, for some values, calculations were performed with averaging of 100 datasets, but no significant difference was observed.

Figure 5.6 shows the results of a numerical experiment to study the influence of the measurement errors of $\mathbf{E}_\theta$ and $\mathbf{E}_\varphi$ in the near-field region on the $\mathbf{E}_\theta$ and $\mathbf{E}_\varphi$ in the far-field region. The experimental design is similar to that for the study of angular errors, except that the values of the input fields $\mathbf{E}_\theta$ and $\mathbf{E}_\varphi$ was multiplied by the sum of unity and noise.

5.8 Kinematic Schemes of the Spherical Near-Field Facilities

The kinematic scheme of the spherical near-field facility should be designed in such a way as to provide the possibility to locate the probe directed to the center of the sphere encompassing AUT at any point on this sphere. There is no principal importance by which movement (antenna or probe) the task is solved. From all the variety of kinematic schemes, let us consider those in which probe remains stationary during the measurement process and the AUT rotates sequentially around two mutually perpendicular axes (in addition, in the process of setting up and adjusting the facility, the probe and the AUT can linearly move relative to each other, and the probe can change its polarization plane position). The choice of these kinematic schemes is explained by the relative simplicity of their implementation and the ability to easily regulate the distance between the AUT and the probe in the process of setting up and adjusting the facility.

Let us divide the spherical near-field facility kinematic scheme into two parts: the antenna rotary support and the probe rotary support. Each of them provides object fastening and all movements (shiftings and rotations) necessary in the process of tuning and measuring. In addition, let us introduce name "rotary positioner" for rotary support element, consisting of an electric motor and a faceplate mounted on its axis. This element provides rotation for object mounted on faceplate about one axis. Obviously, at least two rotary positioners are included in the AUT rotary support and one in the probe rotary support.

In general case, four Cartesian coordinate systems and four rotation angles of these systems should be introduced for the analysis of the spherical facility kinematic schemes (Fig. 5.7).

Cartesian coordinate systems:

XYZ—chamber fixed Cartesian system, "rigidly coupled" with the chamber in which the facility is installed. The XOZ plane of this system is parallel to the horizontal floor of the chamber; the XOY and YOZ planes are parallel to vertical chamber walls.

xyz—Cartesian coordinate system of the AUT rotary support—coordinate system "rigidly coupled" with rotary support's faceplate on which the AUT is mounted. The oz-axis of this system is perpendicular to the faceplate, and the axes ox and oy are parallel to faceplate's reference plane. During measurement process, this coordinate system can rotate about axes parallel to or coinciding with the axes of the fixed coordinate system XYZ.

$x''y''z''$—Cartesian coordinate system of the AUT—the coordinate system "rigidly coupled" with the AUT. Axis $o''z''$ of that system is directed perpendicular to the AUT

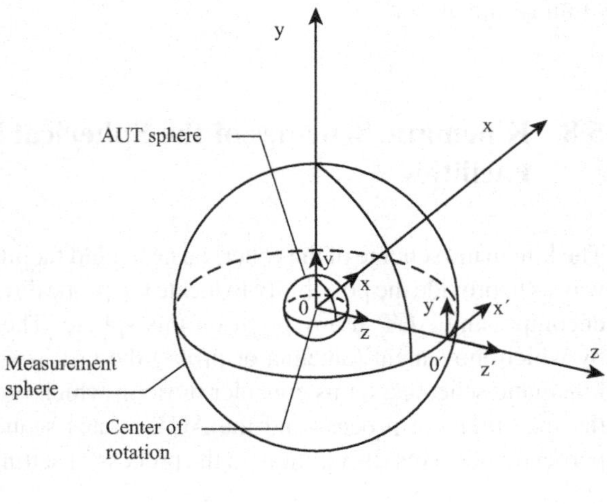

Fig. 5.7 Coordinate systems of: the chamber (*XYZ*), the AUT (matching with AUT rotary support's coordinate system) (*xyz*) ,and the probe (*x′y′z′*). The AUT (*xyz*) and the probe (*x′y′z′*) coordinate systems are in initial position (before any rotation)

aperture (if AUT is an aperture antenna) or in a given direction relative to the antenna. During measurement process, this coordinate system repeats movements of the *xyz* system.

x′y′z′—Cartesian system of the probe rotary support—the coordinate system "rigidly coupled" with the faceplate of the probe. The o′z′ axis of this system is parallel to faceplate normal, and the o′x′ and o′y′ axes are parallel to faceplate reference plane. During measurement process, this coordinate system remains stationary, and during the tuning process, it can rotate around an axis parallel to the normal to the probe rotary support faceplate.

During spherical near-field facility tuning, coordinate systems of AUT (*x″y″z″*) and AUT rotary support (xyz) should be coincided. In case those systems cannot be matched physically, it is always can be done mathematically with the coordinate transform matrices. Therefore, let us leave only three coordinate systems for further analysis: *XYZ*, *x′y′z′* and *xyz*, assuming that *xyz* is the coordinate system of both the AUT and the AUT rotary support.

Angles of rotation of the Cartesian systems' axes:

θ is the angle of rotation of the faceplate of the AUT rotary support around an axis parallel to the oy-axis of the xyz system, or coinciding with it. (azimuth angle).

ψ is the angle of rotation of the faceplate of the AUT rotary support around an axis parallel to the ox-axis of the xyz system, or coinciding with it. (elevation angle).

φ is the angle of rotation of the faceplate of the AUT rotary support around the oz-axis of the xyz system. (roll angle).

χ is the angle of rotation of the faceplate of the probe rotary support around *o′z′* axis of *x′y′z′* coordinate system (polarization angle).

To collect information about the amplitude and phase of tangential component of the electric field on the sphere encompasses AUT, with the probe stationary, it is sufficient to rotate antenna about two mutually perpendicular axes. In principle, following combinations of rotation angles are possible: "θ, ψ," "θ, φ," and "ψ, φ." However, for reasons of ease of implementation, the most common combinations are "θ, ψ" and "θ, φ." They are called: azimuth-over-elevation ("θ, ψ"), elevation-over-azimuth ("ψ, θ"), azimuth-over-roll ("θ, φ"), and roll-over-azimuth ("φ, θ") (Fig. 5.8).

These names carry the following meaning:

- Azimuth-over-elevation (θ, ψ)—the AUT is mounted on the faceplate of the azimuth rotary positioner, which, in turn, is mounted on the faceplate of the elevation rotary positioner (Fig. 5.8a).
- Elevation-over-azimuth (ψ, θ)—the AUT is mounted on the faceplate of the elevation rotary positioner, which, in turn, is mounted on the faceplate of the azimuth rotary positioner (Fig. 5.8b).
- Roll-over-azimuth (φ, θ)—the AUT is mounted on the faceplate of the roll rotary positioner, which, in turn, is mounted on the faceplate of the azimuth rotary positioner (Fig. 5.8c).

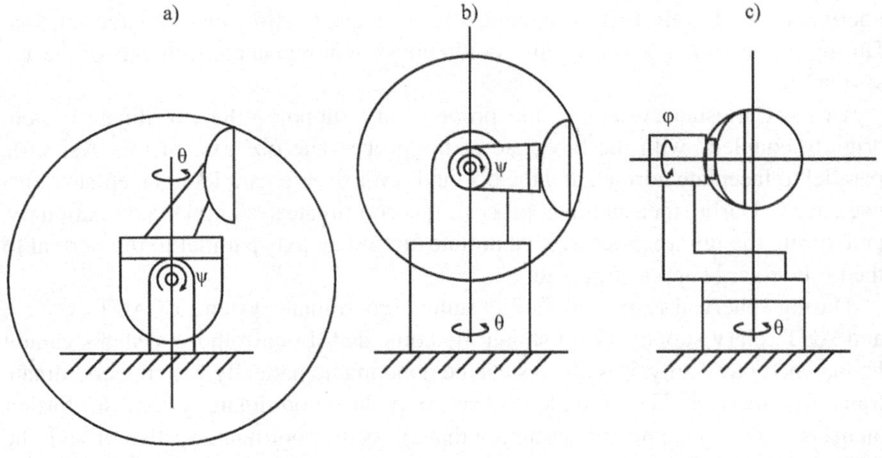

Fig. 5.8 Relative sizes of the minimum sphere encompassing AUT (AUT sphere) for three types of kinematic schemes. **a** "azimuth-over-elevation"; **b** "elevation-over-azimuth"; **c** "roll-over-azimuth"

- Azimuth-over-roll (θ, φ)—the AUT is mounted on the faceplate of azimuth rotary positioner, which, in turn, is mounted on the faceplate of the roll rotary positioner.

Of these four options, the latter ("azimuth-over-roll") is difficult to implement and practically not used.

Let us carry out a comparative analysis of the first, second, and third kinematic schemes of the AUT rotary support implementing the above rotations. The criterion for comparison is the size of the minimum sphere centered at the point of intersection of the rotation axes, into which the AUT fits (AUT sphere). The visual results of this analysis are shown in Fig. 5.8. Obviously, that given same conditions (AUT size, mass, height of the AUT under the facility base) minimum AUT sphere (and, therefore, the minimal influence of reflections) is provided by the roll-over-azimuth kinematic scheme. The basic parts of the AUT rotary support implementing this kinematic scheme are shown in Fig. 5.9. The roll-over-azimuth kinematic scheme is most often used one for spherical near-field facilities.

1. Roll rotary positioner;
2. Slidable tower;
3. Guide-rails and tower linear displacement mechanism;
4. Azimuth rotary positioner;
5. Guide-rails and device four linear displacement mechanism.

The formation of the measurement sphere for the roll-over-azimuth kinematic scheme with discrete roll (φ) values and azimuth (θ) rotation in the range from 0° to 360° is shown in Fig. 5.10.

The formation of the measurement sphere for the roll-over-azimuth kinematic scheme with discrete azimuth (θ) values and roll (φ) rotation in the range from 0° to 360° is shown in Fig. 5.11.

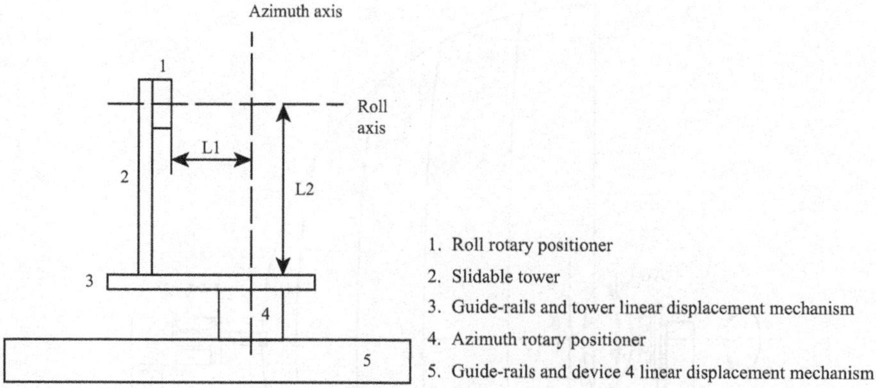

Fig. 5.9 Basic parts of the AUT rotary support for the roll-over-azimuth kinematic scheme

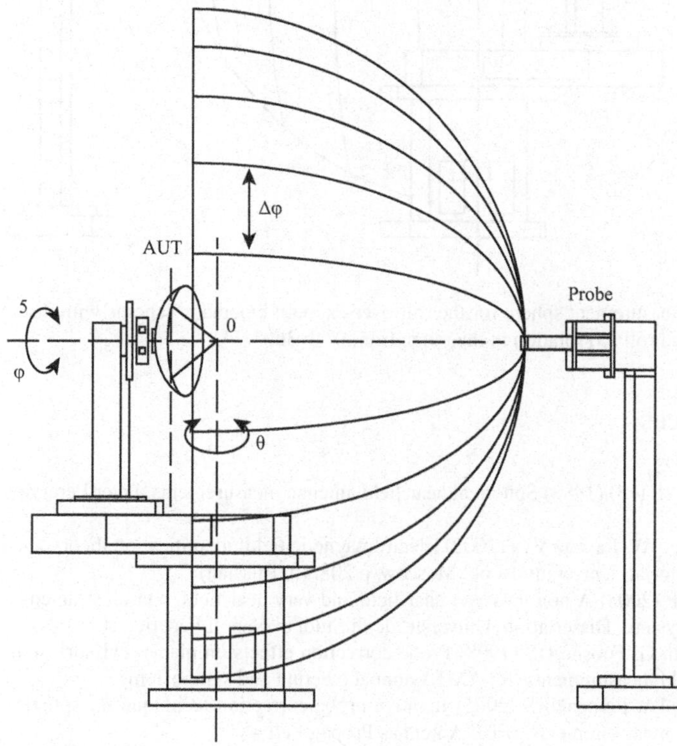

Fig. 5.10 Measurement sphere for the roll-over-azimuth kinematic scheme with discrete roll (φ) values and azimuth (θ) rotation from in the range 0° to 360°

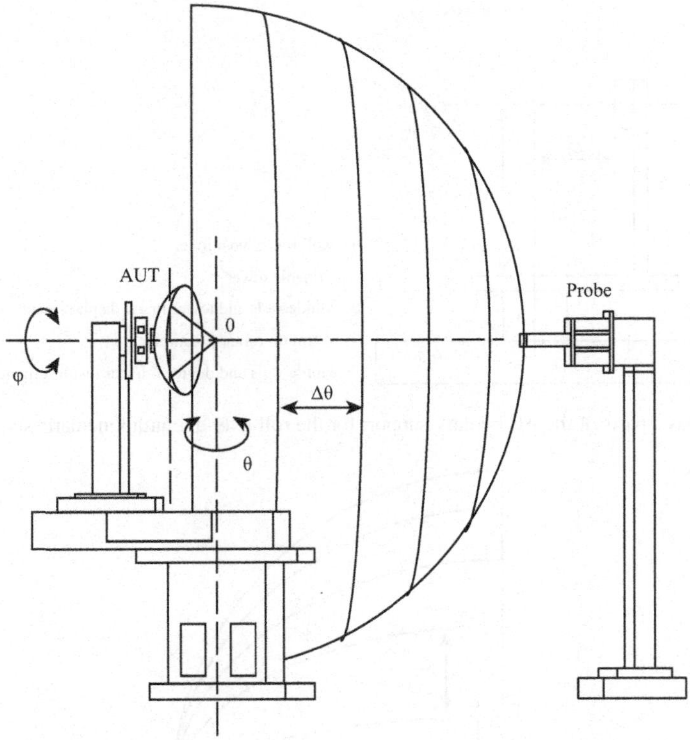

Fig. 5.11 Measurement sphere for the roll-over-azimuth kinematic scheme with discrete azimuth (θ) values and roll (φ) rotation in the range from 0° to 360°

References

1. Hansen JE (ed) (1988) Spherical near-field antenna measurements. Peter Peregrinus, London, UK
2. Kirpanev AV, Lavrov Vya (2002) Electromagnetic field: identification theory and its application. Tutorial. University Book, Moscow, p 208. (in Russian)
3. Vinetti P (2008) A non invasive, near-field and very near-field, phaseless antenna characterization system. Dissertation. Universit‘ degli Studi di Napoli Federico II
4. Hindman G, Fooshe DS (1998) Probe correction effects on planar, cylindrical and spherical near-field measurements. In: AMTA annual meeting and symposium
5. Latinen TA, Pivnenko S (2005) Iterative probe correction technique for spherical near-field antenna meausurements. IEEE Antennas Propag Lett 4
6. Naghski DH, Boyd JT, Howard E et al (1994) An integrated photonic Mach-Zehnder interferometer with no electrodes for sensing electric fields. J Lightwave Technol 12(6):1092–1098
7. Hirose M, Kurokawa S, Ameya M, Komiyama K, Hindawi Publishing Corporation (2012) Near-field anteena measurements using photonic sensor of Mach-Zehnder interferometer. Int J Antennas Propog 2012:11. Article ID 470541. https://doi.org/10.1155/2012/470541
8. Nagatsuma T (2004) Characterization of millimeter-wave antenna using photonic measurement techniques. Int J RF Microwave Comput Aided Eng 14(3):290–297
9. Guzeev IV, Indenbom MV (2006) Restoring the excitation coefficients of the emitters of a linear antenna array and its radiation pattern in the far zone from the measured meridional

radiation pattern in the Fresnel zone. In: Collection of reports of the XVII scientific and technical conference VNIIRT. Radar Systems and Technologies, VNIIRT, Moscow
10. Indenbom MV (2013) Restoring the directivity pattern of a linear antenna and eliminating the influence of reflections from measurements in the Fresnel zone. Antennas (3):190
11. Serhir M, Geffrin JM, Litman A, Besnier P (2010) Aperture antenna modeling by a finite number of elemental dipoles from spherical field measurements. IEEE Trans Antennas Propag 58(4):1260–1268. (Article ID 5398845)
12. Petre P, Sarkar TK (1992) Planar near-field to far-field transformation using an equivalent magnetic current approach. IEEE Trans Antennas Propag 40(11):1348–1356
13. Joy EB, Dingsor AD (1979) Computer simulation of cylindrical surface near-field measurement system errors, CH1456–3/79/0000–0565. School of Electrical Engineering Georgia Institute of Technology, Atlanta, Georgia 30332
14. Rutkowski K, Iversen P Methods to develop complete and accurate error budgets for antenna measurements
15. Newell AC (1988) Error analysis techniques for planar near-field measurements. IEEE Trans Antennas Propog AP-36(6)

Chapter 6
Antenna Testing

6.1 Tests Purpose and Objective

Antenna systems operational features define their testing scope in the course of development. The tests can be split into laboratory (climatic and mechanical) and flight ones. The purpose of all types of tests is to confirm the performance of the characteristics specified in the course of development and maintaining of these characteristics of antenna systems under operating conditions. Testing of antenna systems, apart from flight tests, is normally carried out by the developers within a long time period. The test programs always provide for such types of tests that would be equivalent to the duration of the entire life cycle of antenna systems [1]. As opposed to flight tests of on-board equipment, which can be carried out on special aircraft, flight tests of antenna systems are always carried out on the same aircraft for which they are designed. Depending on the purpose of the aircraft, it is not always possible in the course of tests performance to install on-board instruments, recording antennas characteristics. Therefore, the test program should provide an opportunity to indirectly confirm the performance of those characteristics that are required to be identified in the course of flight experiments.

6.2 Laboratory Tests

Aircraft on-board antennas and on-board antenna-feeder systems (AFSs) are to the greatest extent, compared to other types of on-board radio equipment (e.g., receivers, transmitters, etc.), exposed to destabilizing environmental factors. Laboratory testing of on-board antennas and antenna-feeder systems is always provided by relevant regulatory and technical documentation (technical requirements for the development, technical conditions for the developed and manufactured radio equipment, etc.). All on-board antennas shall provide and maintain their characteristics within the limits set forth in this regulatory and technical documentation, before and in the

V. S. Kalashnikov et al., *Near-Field Antenna Measurements*, Springer Aerospace Technology, https://doi.org/10.1007/978-981-33-6436-3_6

course of laboratory testing, as well as after testing. Requirements for external effects (destabilizing factors norms) on the on-board antennas are set depending on their service conditions and are determined by the type of aircraft, antennas and units location and fastening. Naturally, if any antennas can be used with aircraft of various types, then they are subject to the standards for laboratory tests that characterize the most adverse operating conditions.

On-board AFS normally consists of many units operating in reality in different operating conditions. Therefore, these units are tested separately per external effects corresponding to the operating conditions of particular unit. When testing antennas, special attention should be paid to avoidance of electromagnetic fields, which may cause test equipment malfunction [2]. Mechanical and climatic testings of antennas and AFS are carried out after inspection for compliance with the design requirements specified in the technical specifications, namely compliance with the drawing, coatings quality, etc., as well as radio-technical and electrical requirements [3]. These inspections are usually carried out under normal climatic conditions, unless otherwise specified in the technical specifications for the antenna or AFS. Normal climatic conditions:

- ambient temperature from +15 to +35 °C;
- relative humidity from 45 to 80%;
- atmospheric pressure from 630 to 800 mm Hg.

In compliance with the regulatory and technical documents, the following parameters of aircraft on-board antennas and on-board AFS are usually monitored during laboratory tests [4]:

- VSWR measured at antenna RF output or on-board AFS channel (i.e., including feeder–RF interconnects, cables and feeder units);
- antenna radiation pattern;
- antenna gain or AFS channel transmission factor;
- antennas polarization characteristics, including cross-polarization suppression for on-board radial system equipment antennas.

The measurement methods used in the tests are widely known to specialists. Measurement of gain and cross-polarization suppression of on-board landing and navigation antennas presents certain difficulties, especially in the meter wave band. In cases where these measurements are not carried out in anechoic chamber, but at open testing grounds, measures should be taken to exclude underlying surface effect on the measurement result, or this effect should be taken into account in the course of measurements. The latter is usually implemented by introduction of a correction factor into the resulting expression for parameters calculation on the basis of some correction factor measurement data, which is also determined by measurement in the course of measuring unit calibration. Elimination of underlying surface effect on the gain measurements result is also possible by averaging the measured values of the parameter when changing antenna suspension height.

6.3 Mechanical Tests

When carrying out mechanical tests of on-board antennas and on-board AFS (per unit), it is recommended to install the units on vibration or shock testing machines in the position in which they are installed on the aircraft. This is done in order to bring the test conditions as close as possible to the actual operating conditions. From this point of view, it is also mandatory to use real attachment when placing the units directly at the machines or at the transition elements, repeating the real installation area design and units attachment fittings. Mechanical tests are split into two groups. The purpose of the first group test is to check maintaining of normal operation of the antenna in case of physical impact on it; the purpose of the second group test is to verify the on-board antenna ability to withstand the destructive effect of physical impact and maintain normal operability of the antenna after physical impacts. Naturally, in the course of mechanical tests, special attention is paid to the control of the absence of any mechanical damage to the tested antennas (units). The first group tests include:

- vibration-resistance test;
- impact-resistance test.

The second group tests include:

- structural parts resonance stability test;
- test for resistance to vibration in the frequency range;
- vibration test at one frequency (in-process test, 100% of manufactured antennas are subject to this test in order to detect errors in parts manufacturing and antennas assembling);
- impact-resistance test;
- linear and angular acceleration effect test.

On-board antenna is considered to have passed the second group test, in case of lack of structural changes after each test, and specified antenna parameters meet the technical specification requirements. The first group tests necessitate, as noted, to monitor the on-board antenna performance directly in the course of testing.

Figures 6.1, 6.2, 6.3, and 6.4 contain circuits of some measuring units for such monitoring.

Facility shown in Fig. 6.1 contains the generator and indicator parts. The generator part is formed by transmitting antenna, continuous attenuator, and generator controlled by output power meter. The indicator part includes the tested on-board antenna, installed at the test bench, and the display unit (receiver with display). Prior to tests beginning, the measuring unit shall be calibrated so that the indicating instrument displays value convenient for monitoring. This is achieved by changing the sensitivity of the measuring receiver or changing the emitted signal power using the continuous attenuator. Transmitting antenna shall be installed so that it is no obstructions in front of it, if possible, and at a height, typically approximately equal to tested antenna position. During the test, any movements, other than that of the tested antenna, of any objects or components of the measuring unit are not permitted.

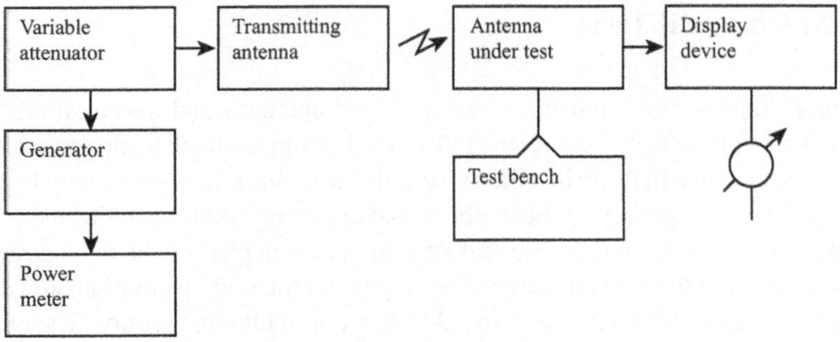

Fig. 6.1 Facility for antenna test using a transmitter

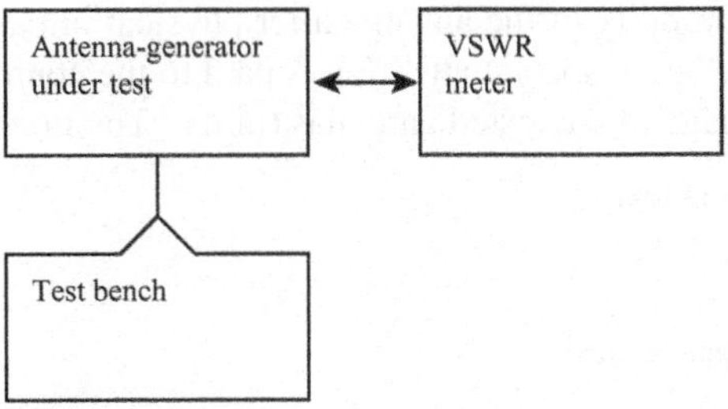

Fig. 6.2 Antenna test circuit using standing wave meter

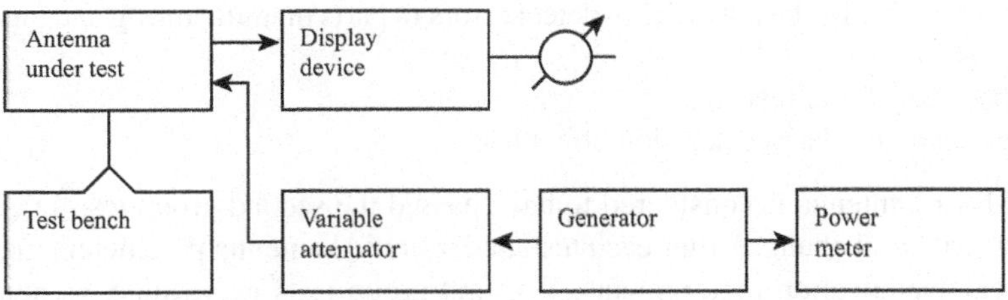

Fig. 6.3 Monitoring circuit for on-board antennas with two outputs

Throughout the test, the radiated power shall be monitored as compared with the original power meter reading obtained prior to the test commencement (during calibration) in order to avoid the power change effect on the measurement process. From the same point of view, any adjustments of the measuring facility after calibration are not permitted. The considered signal passage measuring facility can be used only for testing of antennas with operating wavelengths many times exceeding the normalized amplitudes of test benches movement. In cases when this condition is not met, there is a sharp change in the passing signal value with the bench movement, due to significant change in the phase and amplitude correlation of signals coming to the receiving point from the transmitting antenna with direct and reflected rays. In such

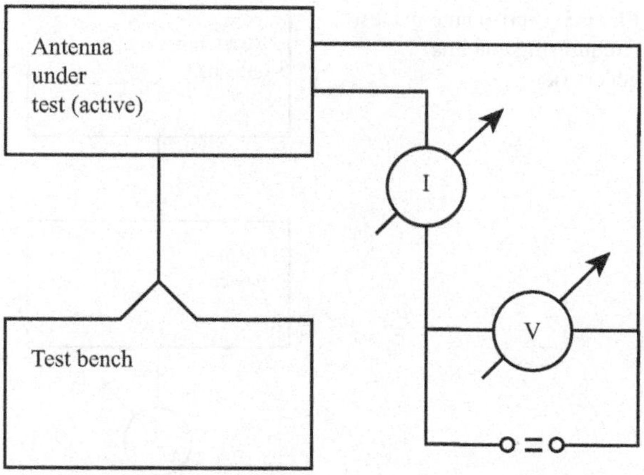

Fig. 6.4 Circuit for active antennas testing on the bench

cases, it is possible to use a conventional circuit to measure the standing wave ratio to monitor antenna operating capability during mechanical tests, as shown in Fig. 6.2.

When using this circuit, it is necessary to pay attention to the soundness of high-frequency connections between the units, and elimination of mechanical impact transfer from the bench through high-frequency feeder or other communication lines (e.g., power wires, etc.) to other elements of the measuring facility, and thereby preventing possible errors in testing. Despite the simplicity and user-friendliness of the measuring facility per circuit of Fig. 6.2, preference within the specified limits should always be given to the first monitoring circuit (Fig. 6.1) for signal indication, as this circuit is complete equivalent of actual communication line. In some cases, it is possible to install both tested and transmitting antenna at the bench. This is particularly preferable, in case another tested antenna is used as the transmitting one. For on-board antennas with two outputs, the signal passage monitoring circuit is quite **significantly** simplified. Measuring unit circuit for this type on-board antennas is shown in Fig. 6.3.

Here, the generator part of the circuit is similar to the circuit of Fig. 6.1, connected to one of the high-frequency outputs of the antenna, and indicator part of the circuit is connected to the other output. Note that the circuit shown in Fig. 6.3 is appropriate to be used only for such antennas, two outputs of which are formed, for example, by the use of two exciters, but not by the use in the antenna design of any separation feeder circuit (power divider, bridge, etc.). In the latter case, application of measuring unit per Fig. 6.3 circuit may result in insufficient coverage of tested antenna functional components. One of the options for monitoring antennas operability in the process of mechanical tests is represented by the circuit, shown in Fig. 6.4.

The control method consists in registration prior to test start and monitoring of parameters of active element circuits power supply circuit in the course of the test. For instance, role of such an active element can play a transistor in active antenna. Installation per Fig. 6.4 at the same time may be additional when testing the receiving active antenna, in case the main performance monitoring is carried out with one of the circuits of Figs. 6.1, 6.2, and 6.3. For testing of active transmitting antennas, it

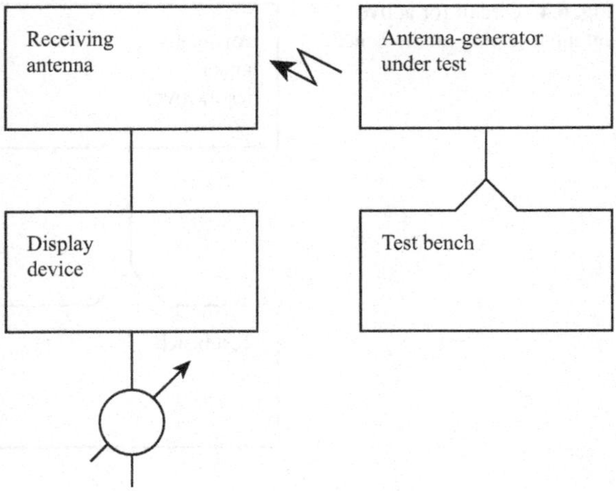

Fig. 6.5 Active antenna test circuit using antenna generator

is possible to use the measuring facility, the circuit of which is shown in Fig. 6.5 (for antenna generator case). In case of mechanical impact tests with operability verification, parameters control shall be carried out in compliance with technical specification norms for on-board antenna or AFS.

6.4 Climatic Tests

Similar to mechanical, climatic tests of airborne antennas are recommended to be performed with installation in the test chambers in the position, as they are installed on the aircraft. Climatic tests can be conventionally split into two groups. First group tests are intended to test the possibility of on-board antenna normal operability directly exposed to destabilizing factors. Second group tests are intended to establish antenna ability to offset destabilizing factors impact and to maintain antenna operability after climatic effects. The first group tests include:

- heat resistance test;
- cold resistance test;
- frost and dew impact test;
- low pressure test (for transmitting antennas).

The second group tests include:

- temperature cycling test;
- moisture resistance test;
- salt fog test;
- dust-proof test;
- solar radiation exposure test.

The listed types of tests do not exclude performance of additional climatic tests due to relevant operating conditions. When conducting the first group climatic tests, antenna operability control in the course of testing may be performed using one of the reviewed measuring facility circuits with the only difference that the tested antenna shall be placed in the corresponding test chamber in front of a dielectric radiotransparent window (for measuring units per circuits Figs. 6.1, 6.5). Testing performance is significantly simplified if the tested antenna has control output (e.g., output of control rod or control coupling loop placed in resonator of main slot radiator). Antenna operability is then monitored by feeding a high-frequency signal directly to the antenna control output without the use of any separate transmitting antennas. Antenna is considered to have passed climatic tests in case it maintains its operability in the course of test performance and after completion of the test. At the same time, both operability control in the course of testing and the control of the antenna's capability to maintain its parameters after exposure to climatic factors shall be carried out according to the standards specified in the technical specifications for the on-board antenna. In general, laboratory, mechanical, and climatic tests are detailed verification of correctness of technical and design solutions incorporated in antenna design.

6.5 Flight Tests

On-board antennas flight tests are carried out to determine the compliance of the parameters with the specified requirements, taking into account the influence of the airframe on the characteristics of the antenna radiation in real flight conditions [3, 5]. Measurements of low-directional on-board antennas radiation pattern are of particular interest: the frame, fixed fin and controllable planes of the aircraft have the strongest effect on directional properties of antenna. Only in-situ measurements enable to obtain radiation pattern of on-board antennas, taking into account such phenomena as resonant elements excitation (slots in the skin, protrusions, pylons, other antennas), surface waves excitation. In the most general form, the methods of flight measurements of specified characteristics are described in the study [5]. Let us review some specific features of these methods on the example of measurement of radiation pattern of aircrafts low directional antennas.

Figure 6.6 contains generalized measurement diagram. Assume that radiation patterns of on-board receiving antenna are measured. Measurements are made as follows. The ground generator (transmitter) (1) emits signals through the auxiliary antenna (2) in the direction of the aircraft, performing maneuvers, in which the antenna under study (3) takes different positions relative to the auxiliary antenna (2). The signals received by the antenna under test (3) are transmitted to the linear measuring receiver (4), the output of which is connected to multichannel aircraft recorder equipment (5). Required radiation pattern measurements, data related to aircraft turns, and spatial position with regard to the ground station (in the absence of telemetry communication channel) are received and registered both on board from

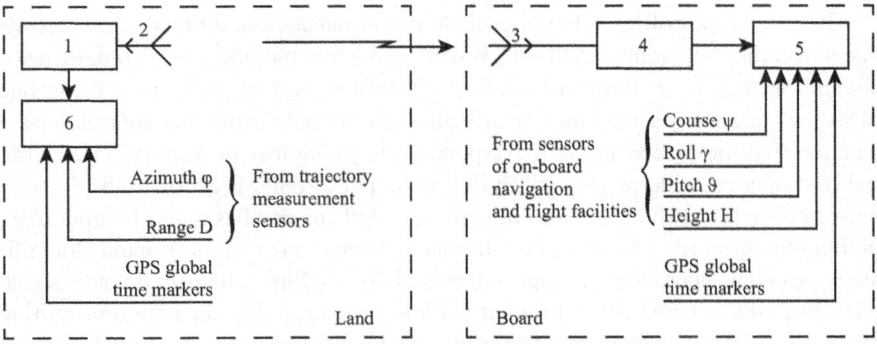

Fig. 6.6 Block diagram of radiation pattern measurement in flight conditions

navigation and flight equipment sensors (course ψ, height H, roll angle γ, pitch angle ϑ) and on earth (aircraft azimuth φ and the distance D relative to auxiliary antenna (2) location). Specified coordinates are received from ground-based means of trajectory measurements, for example radar. To synchronize airborne and ground-based records, GPS global time marks are used. To construct 3D pattern of the on-board antenna, it is necessary to define this antenna gain (or transmission coefficient) variation with respect to direction to the ground station in the coordinate system r, θ, φ associated with the aircraft, for which the following formulas (with angles depicted in Fig. 6.7) can be used [5].

$$\cos\theta = \sin\theta \sin\alpha \sin\gamma - \sin\theta_0 \cos\alpha \cos\gamma \sin\vartheta - \cos\theta_0\gamma \cos\vartheta \tag{6.1}$$

$$tg\phi = \frac{\sin\theta_0 \cos\alpha \sin\gamma \sin\vartheta + \sin\theta_0 \sin\alpha \cos\gamma + \cos\theta_0 \sin\gamma \cos\vartheta}{\cos\theta_0 \sin\vartheta - \sin\theta_0 \cos\alpha \cos\vartheta}. \tag{6.2}$$

Let us determine the requirements to electrical parameters of the main elements of the measuring circuit (Fig. 6.6) on the basis of error analysis. Registration of received signals is carried out using aircraft recorder equipment. In this case, the registered value of mark deviation rate from the initial position is expressed as [2]:

$$e = s_{ARE} k_{MR} P_{REC}, \tag{6.3}$$

where s_{ARE}—calibration function steepness; k_{MR}—gain of the measuring receiver; P_{REC}—power of received signal at the receiver input.

The desired antenna gain can be determined from the Friis transmission formula and expression (6.3) [2]:

$$G_{REC}(\theta, \phi) = \frac{(4\pi)^2 D^2}{\lambda^2} e^{\beta D} \frac{e}{P_{TR} G_{TR} \eta_{TR} \eta_{REC} s k_{MR}}, \tag{6.4}$$

Fig. 6.7 Definitions of the aircraft angles

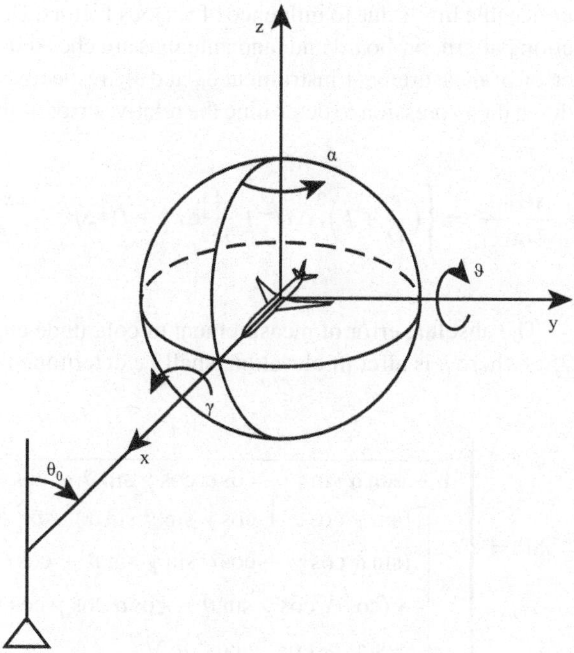

where λ—wavelength, *m*; P_{TR}—transmitter power (pulse), *W*; G_{TR}—transmitting antenna gain; η_{TR}, η_{REC}—efficiency of transmitting and receiving feeders, respectively; β—radio waves attenuation coefficient at the atmosphere, m^{-1}; *D*—range (distance between the aircraft and the auxiliary antenna of the ground station).

It is known that with a significant excess of the measured signal over noise, the absolute error of indirect measurement of some function *f* (*x*, *y*, *z*...) is expressed in terms of absolute errors of mutually independent arguments Δ*x*, Δ*y*, Δ*z* ... in the form of [5]:

$$\Delta f(x, y, z...) = \sqrt{\left(\frac{\partial f}{\partial x}\right)\Delta x^2 + \left(\frac{\partial f}{\partial y}\right)\Delta y^2 + \left(\frac{\partial f}{\partial z}\right)\Delta z^2 + ...}, \quad (6.5)$$

and relative error

$$\delta f = \frac{\Delta f}{f}, \quad (6.6)$$

derivatives $\frac{\partial f}{\partial x}, \frac{\partial f}{\partial y}, \frac{\partial f}{\partial z}$ are calculated at the average arguments values. A number of values included in the expression (6.2), such as efficiency of feeders η_{REC} and η_{TR}, can be determined fairly accurately by direct measurements and are the subject to relatively small changes in the course of testing. The remaining measured values included in (6.4) may change randomly during the flight (and from one flight to another) within noticeable limits due to influence of various factors. Before measurement of the radiation pattern, on-board and land antennas are checked with determination of relative error of measurement instrument δ_B and δ_L, respectively. Given the above, we write down the expression to determine the relative error of the antenna gain measurements

$$\frac{\Delta G_{REC}}{G_{REC}} = \left\{\left(\frac{r}{D} + \beta\right)^2 \Delta D^2 + \frac{4}{\lambda^2}\Delta\lambda^2 + D^2\Delta\beta^2 + \frac{\Delta P_{TR}^2}{P_{TR}^2} + \frac{\Delta G_{TR}^2}{G_{TR}^2} + \delta_L^2 + \delta_B^2\right\}^{1/2} \tag{6.7}$$

The absolute error of measurement of colatitude angle θ at angles $\theta = 90° - \varepsilon \geq 75°$, where ε is aircraft elevation, shall be determined using the expression [1, 5]

$$\Delta\theta = \left\{\begin{array}{l} \dfrac{1}{1-(\sin\alpha\sin\gamma - \cos\alpha\cos\gamma\sin\vartheta - \cos\theta_0\cos\gamma\cos\vartheta)^2} \\ \times\left[\begin{array}{l} (\sin\gamma\cos\alpha + \cos\gamma\sin\vartheta\sin\alpha)^2\Delta\alpha^2 + \\ (\sin\alpha\cos\gamma + \cos\alpha\sin\gamma\sin\vartheta - \cos\theta_0\sin\gamma\cos\vartheta)^2\Delta\gamma^2 + \\ +(\cos\theta_0\cos\gamma\sin\vartheta - \cos\alpha\cos\gamma\cos\vartheta)^2\Delta\vartheta^2 + \\ +\sin^2\theta_0\cos^2\gamma\cos^2\vartheta\,\Delta\theta_0^2 \end{array}\right] \end{array}\right\}^{1/2} \tag{6.8}$$

Expression (6.8) can be simplified, if a certain type of the aircraft evolution is specified, for example, sharp turn with roll angle γ. In this case, the pitch angle ϑ is a few degrees. Assuming that in the expression (6.8) $\sin\vartheta = 0$, $\cos\vartheta = 1$, one can get:

$$\Delta\theta^{EVOL} = \left\{\begin{array}{l} \dfrac{1}{1-(\sin\alpha\sin\gamma - \cos\theta_0\cos\gamma)^2} \\ \times\left[\begin{array}{l} \sin^2\gamma\cos^2\alpha\,\Delta\alpha^2 + (\sin\alpha\cos\gamma - \cos\theta_0\sin\gamma)^2\Delta\gamma^2 + \\ +\cos^2\alpha\cos^2\gamma\,\Delta\vartheta^2 + \sin^2\theta_0\cos^2\gamma\,\Delta\theta_0^2 \end{array}\right] \end{array}\right\}^{1/2} \tag{6.9}$$

Absolute error of the azimuth angle φ, when the aircraft turns with roll angle γgiven elevation less than 15° (angle $\theta_0 \geq 75°$) is determined by the expression [5]

$$\Delta\phi^{\mathrm{EVOL}} \approx \frac{1}{\cos^2\alpha - (\sin\alpha\cos\gamma - \cos\theta_0\sin\gamma)^2} \times \left\{ \begin{array}{l} \cos^2\alpha\cos\gamma + (\cos\theta_0\cos\gamma - \sin\alpha\sin\gamma)^2\Delta\gamma^2 + \\ + \cos^2\alpha \left[\begin{array}{l} (\sin\theta_0\sin\gamma - \cos\theta_0\sin\alpha\cos\gamma) \\ +(\sin\alpha\cos\gamma + \cos\theta_0\sin\gamma)\cos\theta_0 \end{array} \right] \Delta\theta_0^2 \\ + \left[\cos^2\alpha\sin\theta_0\sin\gamma - \cos\theta_0(\sin\alpha\cos\gamma + \cos\theta_0\sin\gamma)\right]^2 \Delta\vartheta^2 \end{array} \right\}^{1/2} \quad (6.10)$$

Note that since the heading angle of the radio station $\alpha = 180° - (\psi + \varphi)$, the absolute error of the angle α is equal to [5]:

$$\Delta\alpha = \Delta\psi + \Delta\varphi = \Delta\psi \quad (6.11)$$

the sum of measurement errors of course angle and azimuth of the aircraft and is approximately equal to the course measurement error, as normally $\Delta\psi >> \Delta\varphi$. The values included in expression (6.7) are determined by:

ΔD—measurement error of the distance between the aircraft and the auxiliary (ground) antenna;

$\Delta\beta$—radio signal fluctuations during propagation in the atmosphere due to instability of its refractive and absorption indexes;

ΔP_{TR}, $\Delta\lambda$—time instability of transmitter power and wavelength;

ΔG_{TR}—fluctuations of the ground antenna gain mainly due to changes in the reflective properties of the underlying surface (ground) and "local objects" depending on the azimuth φ and elevation ε of the aircraft.

Included in the expressions (6.9)–(6.11), absolute measurement errors of corresponding angles are determined by the accuracy of the measuring instruments. The order of ground measurements errors ($\Delta\varphi$, $\Delta\theta_0$) is tenths of a degree, the order of on-board measurements errors ($\Delta\psi$, $\Delta\gamma$, $\Delta\vartheta$)—from fractions of a degree to degree units. From the expressions (6.9) and (6.10), it follows that when the aircraft turns with a given roll angle, the greatest errors in measuring the angles θ and φ should be expected at angles α (and φ) close to ±90° (depending on the sign of the roll angle). Note also that at the course angles of the radio station α = ±90° and the roll angles $\gamma = \arccos\frac{2\cos\theta_0}{1+\cos^2\theta_0}$, the expression (6.9) turns to infinity at finite values of the argument errors. In these cases, it is unacceptable to use the expression (6.9). Angle θ measurement error is uncertain. This is due to uncertainty of the angles φ at the poles of the spherical coordinate system.

Figure 6.8 contains the dependences of maximum absolute errors of measuring angles θ and φ from the angle of roll of the aircraft when performing evolutions. The calculation takes the values of the heading angle of the radio station α = 90°, the measurement error of roll γ, course ψ and pitch ϑ angles.

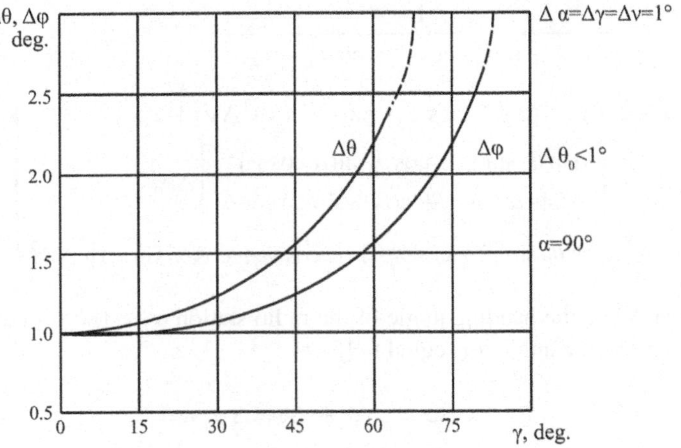

Fig. 6.8 Dependences of measurement errors of angles θ and φ from the angle of roll of the aircraft

$$\Delta\psi = \Delta\psi + \Delta\vartheta \approx 1^\circ. \tag{6.12}$$

errors in measurement of the aircraft azimuth angles φ and angle θ_0.

$$\Delta\varphi = \Delta\theta_0 << 1^\circ. \tag{6.13}$$

From the results obtained above, it follows that the measurement error of angles θ and φ can be several times higher than the absolute measurement error of the angles characterizing the aircraft evolution and position. It should be noted here that in the course of measurements in decimeter and centimeter waves, where the dimensions of the irradiated part of the aircraft are usually hundreds of wavelengths, bursts and troughs width (oscillation period) of the measured radiation pattern is fractions of a degree. Therefore, with the above-specified accuracy of angular measurements, it is not possible to obtain a "thin structure" of the radiation pattern in flight conditions. The smallest details in the radiation pattern of low-directional on-board antennas, obtained by measurements in flight conditions, have a width of the order of several degrees. To improve the accuracy of measurements and reduce the values of the terms in the expressions (6.6) and (6.7), both technical and organizational measures are used. Technical measures include supply voltage stabilization, continuous monitoring, and adjustment of transmitter power and frequency. In general, technical measures are well known, as they are widely used to improve radio communications quality. Let us dwell in more detail on organizational measures that can significantly affect measurements results: choice of place, time, and mode of measurement flight(s). Choice of the place of measuring flights performance of so-called work zone shall be made taking into account the following requirements:

- the work zone shall be located in the "far field";
- steepness of radiation pattern of the auxiliary (ground) antenna 2 (Fig. 6.6) in the work zone shall be minimum both in elevation (vertical), and in azimuthal (horizontal) planes. In decimeter and centimeter bands, this requirement is met, if the ground antenna radiation pattern in the vertical plane has cosecant form and the flights are made at elevations of large five-six degrees and azimuth angles, removed by 5–10° from the directions to notable local items;
- the work zone expansion should be sufficient to "fit" the aircraft with evolutions when performing maneuvers. The turning radius of the aircraft is determined using the formula [2].

$$r_{\mathrm{EVOL}} = \frac{V_{\mathrm{EVOL}}^2}{g \cdot \tan(\gamma)} \tag{6.14}$$

where V_{EVOL}—true speed of the aircraft at a turn, m/s; $g = 9.8$—acceleration of gravity, m/sec^2; γ—roll angle of the aircraft at a turn.

For example, at flight speed $V_{\mathrm{EVOL}} = 167$ m/s (600 km/h) and roll angle $\gamma = 15°$, the turn radius is about 10 km. The height of the flight in the work zone shall be convenient for performing turns with maximum roll angles, i.e., not too high so as not to lose climb power due to air density decrease, but not too low to ensure safe flight. The work zone shall be as close as possible to the airfield to reduce fuel costs on the way to and from the work zone. Precipitation, low stratus, strong wind breaks at altitudes planned for flight can significantly distort the measurement results. The flight mode for measurements is chosen based on the requirements to the spatial sector around the aircraft, where radiation pattern must be known. This sector is determined by limit elevation, roll and pitch angles. At small specified roll and pitch angles, it may be sufficient to measure radiation pattern only in horizontal plane. For that purpose, the flight is made over a given landmark with successive passes over it with changing courses. During such flight, the level of radiation pattern is measured at each course for a long time; the volume of the measurement results is usually sufficient for their statistical processing. At large roll and pitch angles specified for measuring flights of the aircraft, turns of the central area of the work zone with different roll angles shall be performed, as well as climbs, nosediving or "pullups" with different pitch angles. Other aerobatics can also be performed, but only those when the aircraft does not go beyond the boundaries of the work zone, especially vertically. To obtain statistically equivalent results for small and large (steep) evolutions, the latter have to be repeated due to their short duration.

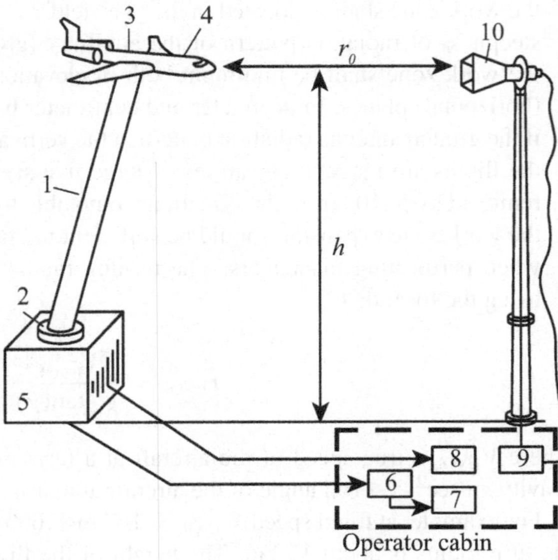

Fig. 6.9 Measuring radiation patterns using an aircraft model

6.6 Methods for Antenna Radiation Pattern Measurement at Model Aircrafts

Let us consider in greater detail the practical methods of large-scale modeling of on-board antennas. When modeling, the operating frequencies of the antenna increase in proportion to the scale factor of the model. The design of the modeling antenna should reproduce the polarization of the studied antenna and its initial radiation patterns. Full and partial modelings are distinguished. In case of full modeling, antenna and the model of the aircraft on which it will be installed shall be performed exactly like the original. In case of partial modeling, the model reproduces only the characteristic details of the original. At that, it is possible to investigate only part of the full-scale antenna parameters. There are many circuits and installation options for measuring 3D pattern of antennas, installed at aircraft scaled models.

One of them is shown in Fig. 6.9, consisting of two towers with tested model and an auxiliary antenna installed at them. Sometimes, a wall or roof of an industrial building is used as a tower for installation of auxiliary narrow-beam antenna. In Fig. 6.10, an example of a measured amplitude–polarization radiation pattern of the antenna and its polarization ellipse in the selected direction is presented.

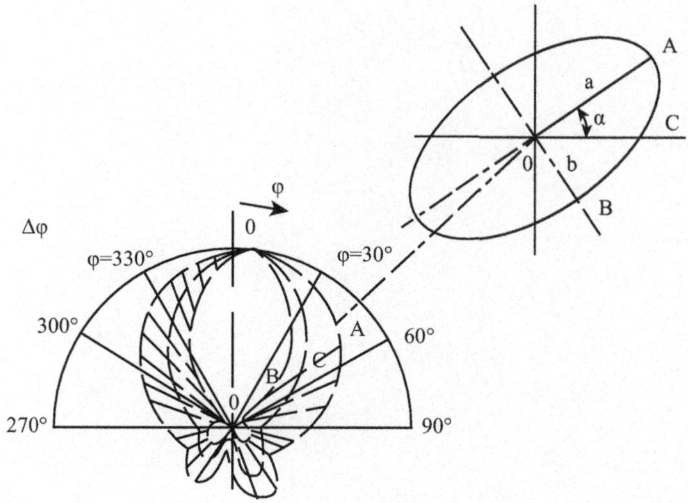

Fig. 6.10 Complex amplitude–polarization radiation pattern

References

1. Shatrakov YuG et al. Aircraft antennas parameters and their measurement. Mashinostroyeniye Publishing House, Moscow, p 208
2. Kazarin AN et al (1971) Methods of calculation and measurement of characteristics and parameters of antennas. Publishing house. BSU after V.I. Lenin, Minsk, p 115
3. Chicago Convention on International Civil Aviation (1947) Annex 10: aeronautical telecommunications
4. Lavrov AS (1974) Antenna-feeder devices. Soviet Radio Publishing House, Moscow, p 367
5. Shatrakov YuG, Rivkin MI, Tsybaev BG (1979) Aircraft antenna systems. Mashinostroyeniye Publishing House, Moscow, p 184

The manufacturer's authorised representative in the EU is Springer Nature Customer Service Centre GmbH, Europaplatz 3, 69115 Heidelberg, Germany. If you have any concerns regarding our products, please contact ProductSafety@springernature.com

Printed and bound by CPI Group (UK) Ltd, Croydon, CR0 4YY
30/09/2026
02225306-0010